G000165245

MASKELYNE

To the students of Latymer Upper

Caitlin [illegible]

MASKELYNE
ASTRONOMER ROYAL

EDITED BY
REBEKAH HIGGITT

ROBERT HALE · LONDON

First published in Great Britain 2014

Published in association with Royal Museums Greenwich,
the group name for the National Maritime Museum,
Royal Observatory Greenwich, Queen's House and *Cutty Sark*

www.rmg.co.uk

ISBN 978-0-7198-0912-5

Robert Hale Limited
Clerkenwell House
Clerkenwell Green
London EC1R 0HT

www.halebooks.com

A catalogue record for this book is available from the British Library

2 4 6 8 10 9 7 5 3 1

Printed in China

CONTENTS

INTRODUCTION

Rebekah Higgitt

THE ASTRONOMER ROYAL

In 1785, the Astronomer Royal, Nevil Maskelyne, sat for a portrait by Louis François Gérard van der Puyl, a Dutch artist working in London at that time (Fig. 1).

Maskelyne was then fifty-three and had been in his post for twenty years. Together the picture and the frame cost £25 14s, with van der Puyl's matching canvas of the sitter's wife and young daughter, painted the following year, costing a further £25 10s.[1] It is an image of a man who has made a success of his life.

Fig. 1: Portrait of Nevil Maskelyne by Louis François Gérard van der Puyl, 1785.

Maskelyne's portrait tells us how he wished to be remembered. He appears in clerical dress, a choice which emphasizes his connections to academia as much as to the church; for, although he was an ordained minister who in 1777 had also been made a Doctor of Divinity, these were the requirements and honours of the University of Cambridge. The year before painting Maskelyne, van der Puyl had similarly portrayed his close colleague Anthony Shepherd, Plumian Professor of Astronomy at Cambridge, also wearing a clerical collar.

In his portrait, Shepherd appears with an orrery, a telescope and a large book, with an assortment of Cambridge buildings visible behind. Maskelyne is likewise shown with appropriate attributes. His left arm rests on the first printed volume of *Greenwich Observations* (1765–74), published in 1776: this had for the first time made the work carried out at the century-old Royal Observatory widely accessible. In the same hand, he holds a diagram of a prismatic micrometer, a device intended to increase the accuracy of measurements made with telescopes that incorporated compound achromatic lenses. This, Maskelyne had invented and, in 1777, the diagram and an accompanying description were published in *Philosophical Transactions*, the journal of the Royal Society of London.

In the background of the portrait is the Royal Observatory itself, within its Greenwich Park setting. The larger building to the right (or west) is Flamsteed House, the home that came with the job and in which he had lived as a bachelor since 1765 and as a married man since 1784. Also just visible to the left are the rooms where the chief instruments were mounted and where the astronomical assistant also slept and worked. This meridian observatory was the modest-looking factory within which regular observations were made and the raw numbers worked, or reduced, into usable data ready for publication.

The portrait was said by Maskelyne's sister Margaret to be a very fine likeness. In the following century another Margaret, the astronomer's daughter, gave it to the Royal Society, where her father had been a Fellow for fifty-three years and, during most of that time, had taken an active role on its Council. Today the painting hangs in the foyer of the Society's premises in Carlton House Terrace.

A LONG LIFE IN BRIEF

Nevil Maskelyne was born in 1732 at the family home in Kensington Gore.[2] He was the third son, and third of four children, born to Elizabeth and Edmund Maskelyne (Fig. 2).

Maskelyne senior was a clerk to the Secretary of State, a post that he held under successive Secretaries of different administrations. The family moved to Tothill Street in Westminster when Nevil was four and the three boys all went on to attend Westminster School, where their father's then-employer, the Duke of Newcastle, was a trustee. This probably helped gain the two older boys scholarships, while Nevil initially attended as a day boy. Their father died in 1744, leaving his family only a small inheritance, although the eldest brother, William, inherited significantly more from his maternal great-uncle. This included the estate of Purton Stoke, where Nevil was later to spend holidays with his wife and daughter.

NATIONAL MARITIME MUSEUM, ZBA5097

Fig. 2: Portrait of Elizabeth Maskelyne, English school, early eighteenth century. The mother of the future Astronomer Royal, Elizabeth Booth married Edmond Maskelyne (1698–1744) in the 1720s. She died in the winter of 1748–9 but this portrait remained with the family.

The Duke of Newcastle's patronage provided for the second Maskelyne son, Edmund, who, on his father's death, obtained the post of writer (a junior servant dealing with minutes, accounts and documents) in the East India Company. He was sent to Madras aged just seventeen but soon settled and, together with a female cousin who was also there, in 1751 persuaded his family to send out his sister, Margaret, to marry his new friend, Robert Clive. The match was made in 1753 and, by the time Margaret returned to England in 1760, Clive 'of India' was rich and famous as a result of his military successes.

Most of what we know of Nevil Maskelyne's schooldays is recorded in his 'Autobiographical Notes', which are discussed more fully in Chapter 1. He said little about his formal classical education, preferring to focus on his extra-curricular interest in astronomy and optics. This developed, he says, as a result of 'occasional discourses in the family' that made him 'eager to see the effects of telescopes on the heavenly bodies, and to know more of the system of the universe'.[3] It is very likely that these family discussions included his sister Margaret; for we know that she too had a lifelong interest in astronomy, as well as music, poetry and languages, and later herself acquired some scientific instruments, including a pair of globes and a telescope (Fig.3 and see Figs 2 and 3 in Case study G, p. 267).

NATIONAL MARITIME MUSEUM, ZBA5113

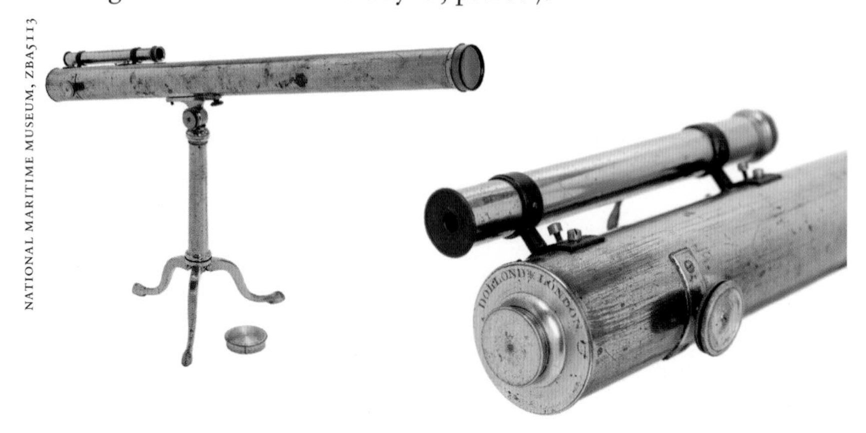

Fig. 3: This early nineteenth century refracting telescope, by Dollond, belonged to the family and has been associated with Maskelyne's sister Margaret, Lady Clive, who in later life had a collection of scientific instruments.

In addition to having a family sympathetic to scientific pursuits, being brought up in Westminster had its advantages: in the eighteenth century, neighbouring London was the world centre of the scientific instrument trade. Thus Maskelyne described how he was able to observe the solar eclipse of 1748 with a camera obscura belonging to a Mr Ayscough. This was undoubtedly the optician and instrument maker James Ayscough, whose shop was at the sign of the Golden Spectacles in Ludgate Street. One of his books on optics and vision included a diagram of a camera obscura, which, he explained, was the most effective means by which to carry out the experiments he described (Fig. 4).

The eclipse, Maskelyne later claimed, gave him the spur to return to his optical, astronomical and mathematical studies with more seriousness. Nevertheless, his formal schooling continued

Fig. 4: The trade card of the optician and instrument maker James Ayscough, c.1750. In the centre is Ayscough's shop sign, the golden spectacles, surrounded by a range of the optical and navigational instruments, including a microscope, telescope and octant.

and, on the death of his mother in 1749, he became a boarder in Vincent Bourne's house in Great College Street. It was his scientific interests that 'led him naturally to the University of Cambridge', he said, although he was following in the footsteps of his eldest brother, who was a Fellow of Trinity College and an unsuccessful candidate for the professorship of Hebrew before retiring to his inherited country estate.

William Maskelyne was still in Cambridge when his younger brother entered Catharine Hall in 1749. Nevil was a sizar, which meant that he paid reduced fees in exchange for performing various menial tasks within college. A notebook containing a list of 'Books bought or given me since my first coming to the University', including standard works on mathematics, magnetism, microscopy and astronomy, gives some record of his course of study, as well as an early example of his lifelong impulse to list and document.[4] For some reason he moved college twice; first to Pembroke, where he matriculated in 1750, and then to Trinity College, where he became a pensioner in 1752, a scholar in 1754 and came seventh in the Mathematical Tripos of that year. He succeeded in being elected a Fellow of Trinity in 1756, having passed a further college examination and taken holy orders.

The Fellowship gave Maskelyne a small stipend and free board and lodging until he married. In 1755 he had also become curate at Chipping Barnet, although we have little evidence of his clerical activities beyond having three times officiated at marriages between 1756 and 1763. However, he was clearly far more interested in making acquaintances and moving ahead within scientific circles. Being a Fellow at Trinity undoubtedly eased this path, for on 27 April 1758 he was elected Fellow of the Royal Society, with the backing of some significant figures. As well as being proposed by Richard Hassell, a fellow Cambridge graduate with whom he lodged and carried out some meteorological observations in Chipping Barnet, Maskelyne was supported by two professors of mathematics at military and naval schools, a teacher of navigation, and James Bradley (1692–1762), since 1742 the third Astronomer Royal at Greenwich.

Maskelyne already had a reputation as an able mathematician and was clearly known within the circles of practical mathematicians. One of his proposers, Thomas Simpson (1710–61), professor of mathematics at the Royal Military Academy at Woolwich, apparently put him first in a list that he shared with the Swedish astronomer, Bengt Ferner, of the best English mathematicians.[5] How this Cambridge mathematics graduate gained a reputation for skill in practical astronomy and instruments is, perhaps, less obvious. He may have had the opportunity to handle astronomical instruments when at Cambridge, where there were small observatories in St John's and Trinity Colleges. He was also close to the physician John Bevis, who had a private observatory in Stoke Newington, only about eight miles from Chipping Barnet. Making acquaintance with Bradley must have been particularly important although, sadly, we know little about how this happened. It has been suggested that Maskelyne assisted with computing work on Bradley's tables of atmospheric refraction in the 1750s, but we do not know who might have recommended him, or when.

It was through such acquaintances, and whatever experience they helped him to gain, that Maskelyne was soon afterwards able to make himself indispensable to the execution of a significant venture for the Royal Society. This was observation of the 1761 transit of Venus, for which the Society was to send astronomers to the Atlantic island of St Helena and to Bencoolen in Sumatra, with the assistance of the East India Company, the Royal Navy and, just before his death, King George II. Bradley had been charged with providing a costed list of necessary instruments, a task which he delegated to Maskelyne. Often adept at taking opportunities when they were presented, Maskelyne had already suggested that whoever was sent to St Helena should make the best of their temporary location by extending their stay and attempting a number of other important observations. Either he or Bradley also realized that the voyage itself could be used for making practical ship-board trials of the lunar-distance method of finding longitude, or position east-west of a given location (see Case study A). It is perhaps not surprising that, after this,

Maskelyne was chosen as one of the observers of the transit of Venus sent to St Helena.

On the voyage and through the programme of observation undertaken during his stay on the island, Maskelyne demonstrated his active involvement in the most significant projects of European practical astronomy in the eighteenth century (see Case study B). These included programmes of observation designed to confirm and extend the work of Isaac Newton, as laid out in his *Principia Mathematica* and subsequently improved by the application of Continental mathematics. These projects could and did take advantage of state interest in the potential for astronomy to aid practical navigation – seen as key to advancing foreign trade and imperial expansion. This gave astronomers both a possible source of patronage and the chance to make observations in different locations around the world, something essential to the success of scientific projects designed to establish nothing less than the shape, scale and motions of the Earth and the solar system.

While cloud-cover at St Helena frustrated Maskelyne's observation of the transit of Venus there, the long stay and two-way voyage gave him opportunity to show his ability to undertake astronomical observations at sea and ashore. He could use this experience to claim a mastery of observing instruments and their proper management and, in addition, he and his assistant demonstrated the potential of the longitude-finding method that they had trialled. All of these things set him up for employment by the Board of Longitude, which appointed him to carry out observations during a voyage to and on the island of Barbados during the official trial of three contenders for its financial rewards in 1763–4. They also made him an authority to whom the Royal Society would listen on the issue of management of the Royal Observatory and, when the position of Astronomer Royal became vacant, a credible candidate to replace Nathaniel Bliss (1700–64), who had held the post since 1762.

The Royal Observatory at Greenwich had been founded in 1675 to improve astronomy for the purpose of aiding navigation. Developing ways of finding longitude at sea, pursued via astronomical and other complementary methods (principally the use of a timekeeper that could keep time on board ship), was to be the most significant work of Maskelyne's life as fifth Astronomer Royal and, *ex officio*, a Commissioner of Longitude. It was work that saw him restructure the regime at the Observatory, undertake the testing of timekeepers, assess a large number of plausible and not-so-plausible ideas submitted to the Board of Longitude and oversee the use of instruments and the making of observations by individuals on voyages of exploration around the world. It also saw him embroiled in more than one controversy with those who disagreed with his judgements.

After his own early voyages, Maskelyne was for the rest of his life settled at Greenwich. He undertook just one more expedition, this time neither maritime nor foreign. Travelling to Scotland, he carried out observations of the effects of the Earth's gravity and, specifically, a phenomenon called 'the attraction of mountains'. The fieldwork was done either side of Schiehallion, a mountain in Perthshire, and its conclusions earned him the Royal Society's prestigious Copley Medal in 1774 (Fig. 5).

As Astronomer Royal at Greenwich, Maskelyne's most significant legacies were the establishment of a constant regime of core, utilitarian observations and their publication in *Greenwich*

NATIONAL MARITIME MUSEUM, ZBA2361

Fig. 5: Maskelyne's Copley Medal, awarded by the Royal Society in 1775 for his observations at Schiehallion. The President of the Society, Sir John Pringle, noted that Maskelyne had submitted himself to the 'inconveniences' of the Scottish mountainside at the Society's request and, thereby, 'firmly established' the theory of universal gravitation.

Observations, of which he oversaw the first four volumes. In addition, these observations fed into the production of pre-computed predictive tables, published annually from 1767 as the *Nautical Almanac*. This considerably simplified the process of finding longitude by astronomical observations. Through its use by navigators and surveyors, the Greenwich meridian became first the national and then the international Prime Meridian, the location from which all longitude positions are measured. In very different form, the *Nautical Almanac* is still published today.

At Greenwich, Maskelyne was aided by a succession of astronomical assistants. He did not marry until 1784, when he was in his fifties. He, his wife Sophia and their only child, Margaret, lived at the Royal Observatory together until his death in 1811. For over another quarter-century after his marriage, Maskelyne continued to be one of the most significant figures in European astronomy and its allied practical and observational sciences. He is not famous for having made discoveries but his observations, influence, advice and management were crucial in underpinning the work of many others. The evidence suggests that although he made some enemies as a result of his duties as a Commissioner of Longitude, most famously the clockmaker John Harrison and his son William, he was otherwise widely respected and liked. He was a man anxious to do his duty, whether that was in making decisions about spending public money or in aiding those – family, assistants, computers, protégés – for whom he felt responsible, but he was also one who enjoyed the company of like-minded friends and colleagues.

This book arises from a one-day symposium held at the National Maritime Museum (NMM), which has been responsible for the buildings of the Royal Observatory, Greenwich, since it became solely a museum and heritage site in the 1950s. This was held in 2011, the bicentenary of Maskelyne's death and, more cheerfully, the 250th anniversary of his 1761 voyage to St Helena. With Maskelyne's work demonstrating the crossovers between astronomy, navigation and scientific exploration, he is a key figure for the Museum. An additional impetus was given by the coincidence of two other circumstances. One was a significant acquisition of material from Maskelyne's descendants, including papers, books, drawings, paintings, instruments and textiles.[6] The other was the early stages of an academic project, undertaken jointly by the Museum and the University of Cambridge, on the history of the Board of Longitude.[7] Linked to these was the 2013 release of the digitized archives of the Board and many of Maskelyne's other papers already held at Cambridge and Greenwich.

These objects, archival material and projects allow us to explore Maskelyne's work, and his domestic and professional contexts, more deeply than ever before. It is those contexts that form the core subject of several of the essays in this collection. As they highlight, Maskelyne's most significant accomplishment was as a manager and organizer: of other people, of data, of information, of publications and institutions. Being involved in some highly collaborative projects, he is a man who can only be considered alongside others. As discussed in the first chapter, which looks at Maskelyne's posthumous reputation, this 'astronomer in context' is perhaps the most interesting counter-narrative to the one in which Maskelyne has been recalled only as the thorn in the side of John Harrison.

It is apt, therefore, that the second chapter, by Jim Bennett, focuses largely on another career altogether, that of Robert Waddington. As well as giving a full account of an important event in Maskelyne's life, it presents a parallel story that could have been his had he not had the privilege of a university education, the backing of useful allies, a sober life and the ability to pay exceptional attention to detail. In laying out almost all that is known of the otherwise obscure life of Robert Waddington, who shared the transformative voyage to St Helena with Maskelyne, we learn about the range of opportunities available to those with knowledge of and ability in astronomy and mathematics in the eighteenth century. Following this, Nicky Reeves's chapter shows just how effectively Maskelyne created his own opportunities for employment, and established the rigorous regime at Greenwich that he demonstrated was (with potentially embarrassing consequences) previously missing.

As Reeves notes, among other things that Maskelyne developed much more extensively than his immediate predecessors as Astronomer Royal were networks of individuals and correspondence. These form the subject of several other chapters. Through them we gain a sense of how Maskelyne operated, his reputation among colleagues in Britain and overseas and how he dispensed his patronage, matching able individuals with available positions. Mary Croarken, for example, looks in Chapter 4 at the experiences of some of Maskelyne's network of human computers, paid by the Board of Longitude to undertake the mathematical calculations needed to produce the *Nautical Almanac*.

Croarken shows that Maskelyne took an interest in the lives and careers of those whose work he directed, even if sometimes only briefly, and that he would sometimes go to considerable efforts to help when they got into difficulties. The same is true of Maskelyne's relationship with some of the many clockmakers that he encountered as part of his work as a Commissioner of Longitude and within the Royal Observatory. While he was attacked by some clockmakers who believed he judged their work unfairly, Rory McEvoy's chapter shows that others benefited considerably from his patronage and his attempts to negotiate within the Board of Longitude. Reliable timekeepers, as both Reeves and McEvoy show, were an essential part of observatory equipment, as well as being part of the answer to finding longitude at sea. This was not a man with a general prejudice against mechanics and artisans, although he clearly did not have a full appreciation of the care with which the new timekeepers had to be prepared and handled.

Maskelyne's networks reached well beyond the circles of London instrument makers and his computers. Alexi Baker demonstrates how correspondence extended his connections across Britain and the Continent and allowed him to share knowledge with, and offer assistance to, those whom he might never meet. As is also shown in the first chapter, he was exceptionally well regarded by French astronomers, as well as within his closer circle of acquaintance. His relationship with some leading men of science was, at times, more strained, echoing wider debates about the way that institutions should support scientific work and interact with individuals and the State. Caitlin Homes's chapter brings this out by focusing closely on the relationship Maskelyne had with the dominant figure in the Royal Society from the 1770s, Joseph Banks. Nevertheless, it was through such personal relationships and correspondence that much of the business of the Board of Longitude was transacted and, by necessity, they found ways to maintain a civil relationship.

The final chapter, by Amy Miller, puts Maskelyne in his most intimate setting – at home in Flamsteed House with his wife and daughter – and connects this one small world with a broader understanding of domestic life and gender roles in the later eighteenth century. We also catch a glimpse of how the astronomer and habitual list-maker approached the business of household management. The portraits and belongings of these family members help us to understand their world, and are considered here, and within some of the shorter sections (case studies) placed between each chapter. These sections also provide the necessary background to understanding the focus of Maskelyne's endeavours and the institutional settings within which he worked.

All the chapters, case studies and illustrations in this book help to put Maskelyne firmly within the setting of the time and place in which he lived and worked. This not only helps to make sense of an important life but also demonstrates the extent to which his scientific work was, as such endeavour always is, a product of specific circumstances. In addition, we cannot understand the development of past science and technology without paying attention to figures like Maskelyne, even though he did little more glamorous than undertaking many repetitive observations that confirmed, supported and extended the work of others, or managing the work of yet more. Such things are, however, as essential as novel discoveries and the formulation of new theories.

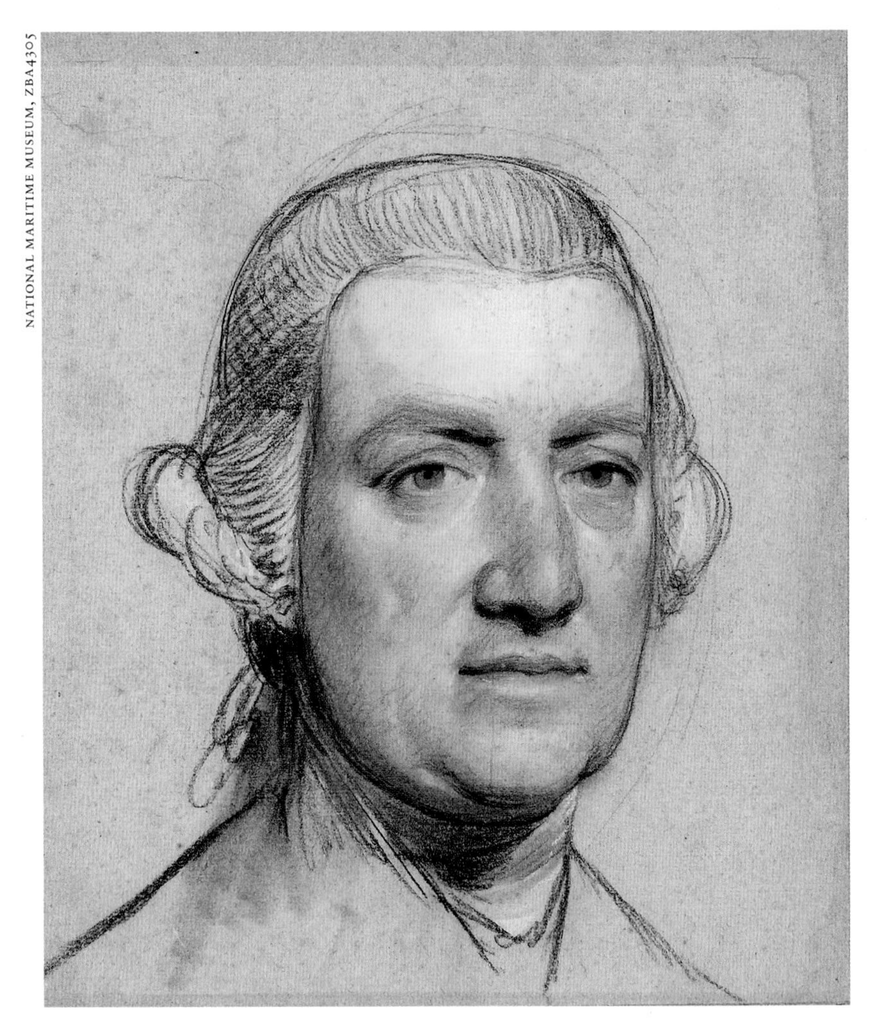

Fig. 6: Portrait of Nevil Maskelyne by John Russell, c.1776. Russell was a leading pastel portraitist and amateur astronomer, noted for his study of the moon's surface. This chalk study of Maskelyne, aged about forty-four, has a striking immediacy.

FURTHER READING

The only full-length biography of Maskelyne is Derek Howse's *Nevil Maskelyne: The Seaman's Astronomer* (Cambridge: Cambridge University Press, 1989). Howse – Head of Navigation and Astronomy at the NMM, 1976–82 – was also author of

'Maskelyne, Nevil (1732–1811)', in the *Oxford Dictionary of National Biography*, Oxford University Press, 2004; online edn, May 2009 [http://www.oxforddnb.com/view/article/18266, accessed 4 Sept 2013]. Maskelyne's sister also has an entry there: H.V. Bowen, 'Clive, Margaret, Lady Clive of Plassey (1735–1817)', *Oxford Dictionary of National Biography*, Oxford University Press, 2004 [http://www.oxforddnb.com/view/article/63502, accessed 4 Sept 2013].

Maskelyne figures largely in histories of the Royal Observatory, Greenwich, the most thorough of which is Eric G. Forbes, A.J. Meadows and Derek Howse, *Greenwich Observatory: The Royal Observatory at Greenwich and Herstmonceux 1675–1975*, 3 vols (London: Taylor & Francis, 1975). Aspects of Maskelyne's work and practice have received thorough attention in Nicky Reeves, 'Constructing an Instrument: Nevil Maskelyne and the Zenith Sector, 1760–1774' (unpublished doctoral thesis, University of Cambridge, 2009). Other useful works include Simon Schaffer, 'Astronomers mark time: discipline and the personal equation', *Science in Context*, 2 (1988), pp. 115–45 and the articles on Maskelyne's astronomical assistants and human computers by Mary Croarken, which are cited at the end of her chapter in this volume.

Beyond scholarship in the history of science, Maskelyne remains of interest to those connected to his family and the places with which he was associated. Published works include Thereza Story Maskelyne, 'Nevil Maskelyne, D.D., F.R.S., Astronomer Royal', *Wiltshire Archaeological and Natural History Magazine*, 29 (1897), pp. 126–37, Mary Arnold-Foster, *Basset Down: An Old Country House* (London: Country Life, 1950) and Brian Freegard, *From Marsh Farm to the Moon: The Maskelyne Family of Purton* (Purton Historical Society, 2012).

In terms of understanding Maskelyne's times and context, the chapters in this book and the works cited in their Further Reading sections are an excellent guide.

1

REVISITING AND REVISING MASKELYNE'S REPUTATION

Rebekah Higgitt

MAKING A REPUTATION

Those who have heard of Nevil Maskelyne today are very likely to have come across him in relation to the clockmaker, John Harrison. The best-known account of Harrison and his sea clocks is undoubtedly Dava Sobel's book, *Longitude: The True Story of a Lone Genius Who Solved the Great Scientific Problem of His Time* (US: 1995, UK: 1996), and the 2000 film based on it.[1] In both, Maskelyne appears as a dull but jealous and snobbish Cambridge-trained cleric, whose elitism and privileging of astronomy over mechanical inventiveness prejudice him against the Yorkshire-born and Lincolnshire-bred Harrison. He is jealous, petty and obstructive, putting potential personal gain over disinterested judgement.

Sobel tells a good tale and, as she writes, 'A story that hails a hero must also hiss at a villain – in this case, the Reverend Nevil Maskelyne'. While she admits that he is 'more an antihero [i.e. a foil to her hero, Harrison] than a villain' and 'probably more hardheaded than hardhearted', she proceeds to tell her story and interpret his actions from the partisan standpoint of Harrison and his son William.[2] Naturally, we get a very different view if we choose instead to listen to Maskelyne and his friends, and to his early biographers, who were sympathetic to their subject. He, of course, also had a life and reputation beyond his acrimonious relationship with Harrison.

Even before *Longitude*, however, the reputations of both men had become intertwined. Their dispute was regularly and perhaps necessarily recalled in nineteenth-century biographies, although without attaching blame to Maskelyne's actions. It was in the twentieth-century rediscovery and championing of Harrison that the latter's increasing dislike of the Astronomer Royal found appeal with a new audience. It is in these more recent accounts, reflecting the disciplinary interests of their writers (and perhaps changing views of social class and the relationship between mechanical and mathematical or theoretical work), that this aspect has come to be seen as a key element of the narrative linking the two men. Such conflict, of course, also heightens the power of a good story.

Rather than aiming to rescue Maskelyne's reputation, or put his version of the dispute with Harrison to the judgement of readers, this chapter will look at how his posthumous reputation has changed over the two centuries since his death. Later ones in this book have stories to tell that will continue to shape our views of Maskelyne, showing his often warm relationships with colleagues, staff, friends and family, and his carefully justified actions as a servant of the state. Here the focus will be on why stereotypes of 'hero', 'villain' and 'maligned reputation' have become dominant yardsticks for considering Maskelyne, although they are clearly problematical in assessing the lives of real people.

DO IT YOURSELF: AUTOBIOGRAPHY

If you want to be recalled in a particular way, the obvious option is to write your own biography or, perhaps, have your reputation safely established through a 'life and letters' assembled by a sympathetic relative. In Maskelyne's case, there was no such substantial account and, in fact, by having chosen not to respond in print to some of his most vocal critics, his voice is missing to a surprising degree from discussions about the controversies in his life. Artisan watchmakers might seem to be at a disadvantage against a member of the university-trained intellectual elite, a cleric and Fellow of the Royal Society, but there is little in the Maskelyne archive or literary canon that competes directly with the pamphlets or 'Journal' assembled by Harrison and his supporters.[3]

Nevertheless, Maskelyne did write some 'Autobiographical Notes', labelled as being 'in Dr. Maskelyne's own hand' by his daughter Margaret, who was to take a role in assisting the appropriate commemoration of her father after his death, but did not go so far as writing his biography herself. The 'Notes' are not a considered memoir, being very patchy in what they cover. Kept among the archives of the Royal Observatory, they were used as a source in various biographies and were first published in full as an appendix to Derek Howse's *The Seaman's Astronomer* in 1989: today they are freely available online.[4]

Although they only treat the period up to the 1780s, Howse dated the 'Notes' to 1800, from content and the fact that the paper is watermarked 1798. He speculated that they could have been written to assist the composition of a biography that appeared in Abraham Rees's multi-volume *Cyclopaedia*. However, that seems unlikely, since it did not appear until 1812 – when Maskelyne was dead, as was usual for biographical subjects in dictionaries and encyclopaedias. We know from correspondence that Margaret Maskelyne forwarded an account of her father's life to the authors of the Rees *Cyclopaedia* entry and of his principal obituary (first published in France), which both clearly show knowledge of a version of the 'Notes', but it seems more likely that Maskelyne wrote them earlier, perhaps with mortality rather than any particular publication in mind. They were not, for example, available to the author of a brief biography that appeared in 1805 in the *European Magazine*, six years before he died.

The 'Notes' are the basic source for Maskelyne's very early life, to which later historical research has made only minor additions. There is, though, precious little information given. We get a simple pedigree – he is 'the last male heir of an antient family' in Wiltshire, of possible Norman ancestry – and gather that, although he received his formal education at Westminster School, he gained 'instructions in writing & arithmetic during intervals of school from other masters'. His interest in 'the two kindred sciences of Optics & Astronomy', which he saw as the chief pursuit of his life, were kindled by 'occasional discourses in the family' and by observing the 1748 eclipse of the Sun with a 'Mr Ayscough' (probably the optician and instrument maker, James Ayscough), projected through a telescope onto a white screen in a darkened room, or camera obscura.

While Maskelyne claimed that he studied mathematics only as a means to his end of pursuing astronomy, he made it clear that he was very good at it. In only 'a few months, without any assistance' he had sufficiently mastered geometry and algebra to read widely in natural philosophy. This led him to a degree at the University of Cambridge. The 'Notes' claim that he was placed third 'on the list of the first rank of honours' while Cambridge records state that he was in fact seventh wrangler, although it is certainly true that he was elected a Fellow of Trinity in 1756 and of the Royal Society in 1758.[5] His account is then frustratingly silent on the period in which he became a curate and made the acquaintance of key figures like James Bradley, which still remains something of a mystery.

After this brief beginning, the 'Notes' quickly move onto the familiar territory of transits, longitude and the Royal Observatory, Greenwich. It is, as one would expect of a biography of the period, focused on public life and scientific achievements rather than matters personal. The pivotal moment of his life is therefore presented as the voyage to St Helena in 1761. While Maskelyne highlighted the disappointments of this 'trying occasion', he credited his fortitude in accepting his misfortune and 'the fertility of his astronomical knowledge & pursuits' in finding a way to turn a disappointment into something of lasting importance. Thus his sea-trials of the lunar-distance method of finding longitude, based on Tobias Mayer's lunar tables, are presented as a resounding success, demonstrating its 'practicability', 'certainty' and 'utility'.

The 'Notes' establish what Maskelyne thought were his most significant achievements. First, of course, came his practical vindication of the lunar-distance method, while the rest all relate to making the Royal Observatory a useful, accountable and efficient institution. He notes his various improvements to the Observatory's instruments, but particularly his work to make its output available to astronomers and navigators. As well as being the first Astronomer Royal to publish observations regularly, Maskelyne was proud of having rescued James Bradley's earlier ones for the nation. Above all he felt pride at having set up the *Nautical Almanac* as a means of transforming observations made at Greenwich into something of public utility, and of fulfilling the Observatory's original purpose of supporting navigation. Stressing that this labour was undertaken 'voluntarily', Maskelyne called the *Almanac* 'a work which every lover of Astronomy, Navigation & Commerce would readily say, with Emphasis, Esto Perpetua!'[6]

Including the fact that his work on the *Almanac* was unpaid was, most probably, a response to the insinuation that his dealings with Harrison and other watchmakers were coloured by self-interest. In other words, he hoped to make it clear that any promotion of the lunar-distance method was not based on hopes of personal financial gain. Despite Harrison having been dead for quarter of a century by the time that Maskelyne penned his 'Notes', arguments with Thomas Mudge and John Arnold and their sons were recent or still ongoing. Maskelyne wanted to do his best to demonstrate that the judgements which he was required to make, as Astronomer Royal and Commissioner of Longitude, were based on his understanding of what would be most beneficial to the nation.

It is telling that the only titled subsection in his manuscript 'Notes', and the one under which he emphasized the important legacy of the *Nautical Almanac*, is on the 'History of Mr. Harrison's time-keeper, & of Mayer's Tables', and events nearly forty years old. It suggests that Maskelyne perceived this issue as still having a particular resonance that would affect the way he would be remembered. Here he briefly explained the Board of Longitude's 1765 response to the trials of Harrison's watch (H4) and the tables. It was not simply his decision, he emphasized: the rewards for both were 'unanimously agreed to' by the Commissioners present. In addition, the judgement of the trial was not just about whether and how much to reward Harrison but also signalled the 'beginning of that highly useful publication, the nautical almanac'.

EARLY BIOGRAPHIES AND OBITUARIES

Although the 'Notes' were probably written a few years previously, it is tempting to see them as a response to a text that might be considered the 'public' view of Maskelyne in the early nineteenth century. In 1805, the *European Magazine* published a very brief account to accompany an engraved portrait of him, based on a pastel drawing of 1804 by John Russell RA (Fig. 1).[7]

Given that this portrait was in Maskelyne's possession, he presumably gave permission for it to be engraved, but it is clear that he had little to do with the accompanying text which, while respectful, noted that a scarcity of available biographical materials meant that there was a limited amount to say.

The public Maskelyne was, according to this brief outline, a 'very worthy divine and profound mathematician' who had produced 'a very useful practical work' in *The British Mariner's Guide*. It otherwise contains broadly what we would expect: his role in the transits of Venus and longitude trials, followed by the long Harrison affair. The fact that decisions made as a Commissioner of Longitude had created enemies was common knowledge, although a footnote states that the 'insinuation of interestedness in Dr. Maskelyne, all who knew him were sure was without any foundation'. The rest of the account is largely based on his published work: lunar tables, which were 'astonishing evidence of painful industry'; his editorship of 'all the very numerous publications issued by the Board of Longitude'; publication of *Greenwich Observations* and his contributions to *Philosophical Transactions*. These last evidently did not hold much interest for

Fig. 1: This engraved portrait of Nevil Maskelyne was published in the *European Magazine* in 1805 to accompany a brief biography. It is based on an 1804 pastel portrait by John Russell (see p. 269).

the author, being simply described as 'not more remarkable for number than for importance'.

The article in Rees's *Cyclopaedia* (1812) was a more substantial biography, which had the benefit of Maskelyne's 'Notes', in whatever form they were transmitted by his daughter, and personal acquaintance with him. The author was the mathematician and astronomer Patrick Kelly (1756–1842), who was Master of the Finsbury Square Academy, a metrologist and examiner for Trinity House.[8] Taking his cue from the 'Notes' and, undoubtedly, his own interests, Kelly praised the *Nautical Almanac* as one of Maskelyne's main achievements; 'a lasting monument of labour and profound learning. It is universally allowed to be the most useful work on practical astronomy ever published.' There is also an interesting emphasis on the great admiration foreign astronomers had for Maskelyne, to which his correspondence and foreign membership of several scientific societies also testifies.

Unlike in the 'Notes', John Harrison is not mentioned by name, but a long section focuses more generally on the 'important and laborious duty' Maskelyne undertook on behalf of the Board of Longitude. In particular, there was the ongoing consideration and investigation of proposed navigational and astronomical methods and devices. While dealing with the steady submission of new ideas was onerous enough, it also put Maskelyne in the awkward position of having to make decisions that affected the livelihoods of skilled individuals as well as handling 'numerous candidates of very slight pretensions, and even visionaries'.

Kelly looked to engage the reader's sympathy for Maskelyne – 'it is easy to conceive how arduous as well as unpleasant such a duty must have been' – in judging the complaints, discontent and appeals to Parliament that arose from some of his decisions. He concludes that, 'whatever difference of opinion might have then existed, time and experience have since fully proved the truth and impartiality of Dr. Maskelyne's decisions.' This way Maskelyne is shown to be, at least in part, a martyr to such disputatious individuals simply by doing his duty in public office. A very similar view arises in an 1824 biography written by a man who knew the problems of Maskelyne's job only too well. This was Thomas

Young, who was then Secretary to the Board of Longitude and responsible for the *Nautical Almanac*, over which he was heavily criticized. In addition to understanding the burdens and conflicts it involved, Young also emphasized how much of it was done by Maskelyne without payment (Young being paid in both of his roles).[9]

In the *Cyclopaedia* biography, the general assessment of Maskelyne's achievement is Kelly's and owes nothing directly to the 'Notes'. First, Maskelyne's role within the broader astronomical community is acknowledged: 'In the history of science, few persons can be mentioned who have contributed more essentially to the diffusion of astronomical knowledge ... and perhaps no man has been so successful in promoting practical astronomy, both by land and sea.' Kelly also highlights the point that, as well as publishing and corresponding widely, Maskelyne was often called upon to give advice to astronomers building or improving their observatories. Maskelyne was, of course, a practical astronomer, and in this focus 'he was eminently successful, particularly in his labours for the longitude, by which he essentially contributed to the advancement of navigation, the prosperity of commerce and the wealth, honour, and power of his country'.

Kelly ends, as is traditional for such biographies, with a view of his subject's private character. It was – again, as would be expected – 'truly estimable': he was 'exemplary in the discharge of every duty', as already evidenced in his work for the Board of Longitude, and 'modest, simple, and unaffected' in his manner. Kelly knew Maskelyne personally and we can probably lend some credence to his comment that 'To strangers he appeared distant, or rather diffident; but among his friends he was cheerful, unreserved, and occasionally convivial.' While the 'occasionally' sounds like damning with exceedingly faint praise, 'convivial' here probably does not mean 'good company', as in general modern use, but something closer to its older definition of being fond of feasting and carousing: something, therefore, not to be overdone.

Kelly was not just associated with Maskelyne through their shared interests but was also one of his close circle of acquaintance. His biography notes not only Maskelyne's wide and international correspondence but his attachment to his 'intimate friends', naming William Herschel, Charles Hutton, William Wollaston, Samuel Horsley and Samuel Vince, among other astronomers and mathematicians. By way of conclusion, the *Cyclopaedia* entry confirms that Maskelyne was, of course, 'a sincere Christian' – although it adds nothing on his official church role – and notes that he was survived by his daughter, 'whose education he superintended with the fondest care'.

Apart from being informed by direct personal knowledge, Kelly's account of Maskelyne was perhaps also biased by friendship and mutual sympathies. However, what was probably the best-known of Maskelyne's early biographies was published by someone who knew him through correspondence and scientific work rather than in person. This was Jean Baptiste Joseph Delambre (1749–1822), who had been director of the Paris Observatory since 1804. His version was written in his official capacity as Permanent Secretary for the Mathematical Sciences within the Académie des Sciences. Since Maskelyne was a foreign member of the institution, he, like all members, became the subject of an éloge on his death. First read at the Académie in January 1813, it was published as a pamphlet and an English translation appeared in the *Philosophical Magazine* that summer.[10]

Delambre had been a great admirer of Maskelyne: back in 1787, a letter from Jérôme Lalande (1732–1807), Professor of Astronomy at the Collège de France, had commended him to Maskelyne, noting that he had praised *Greenwich Observations* and, even that, '*vous êtes pour lui le dieu de l'astronomie*' (Fig. 2).[11]

While Kelly's account had simply highlighted Maskelyne's considerable reputation and correspondence with foreign astronomers, this éloge naturally focused more on the national and international context of his work. In particular, it included discussion of what the British astronomical tradition (largely practical and observational) owed to French mathematics and theoretical astronomy. Delambre was, however, also able to point out that Maskelyne's important achievement of establishing the *Nautical Almanac* had been enabled by practical work on the

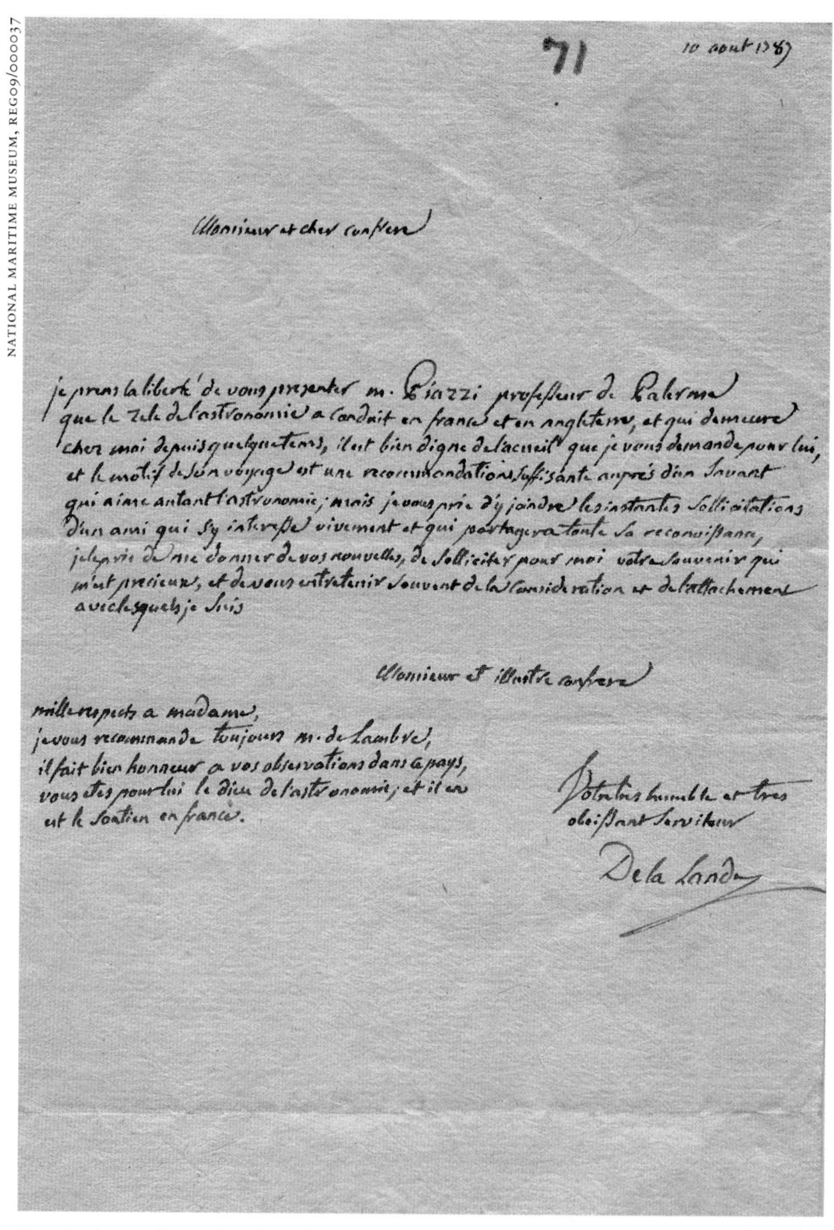

71 10 aout 1787

Monsieur et cher confrere

je prens la liberté de vous presenter m. Piazzi professeur de Palerme
que le zele de l'astronomie a conduit en france et en angleterre, et qui demeure
chez moi depuis quelque tems, il est bien digne de l'accueil que je vous demande pour lui,
et le motif de son voyage est une recommandation suffisante auprès d'un savant
qui aime autant l'astronomie; mais je vous prie d'y joindre les instantes sollicitations
d'un ami qui s'y interesse vivement et qui partagera toute sa reconnoissance,
je le prie de me donner de vos nouvelles, de solliciter pour moi votre souvenir qui
m'est precieux, et de vous entretenir souvent de la consideration et de l'attachement
avec lesquels je suis

Monsieur et illustre confrere

Votre tres humble et tres
obeissant serviteur

De la Lande

mille respects a madame,
je vous recommande toujours m. de Lambre,
il fait bien honneur a vos observations dans ce pays,
vous etes pour lui le dieu de l'astronomie, et il en
est le soutien en france.

Fig. 2: Letter from the French astronomer Jérôme Lalande to Maskelyne, dated 10 August 1787, with a postscript that refers to Jean Baptiste Joseph Delambre's admiration for the Astronomer Royal.

lunar-distance method by Nicolas Louis de Lacaille (1713–62) and directly inspired by his publication of pre-computed tables. However, while it was Lacaille's idea, the English got the credit for making it a regular and useful publication, 'and this is an obligation which seamen and astronomers of all nations and ages have to Dr. Maskelyne'.

As befitted a eulogy, there was high praise for Maskelyne's achievements, particularly his published *Greenwich Observations*. Delambre's assessment of them, as Maskelyne's great legacy to the Greenwich Observatory and British astronomy, has often been quoted in histories of the subject:

> We may say of the four volumes of observations that he has published, that if by any great revolution the sciences were completely lost, and that this collection was preserved, there would be found in it sufficient materials for rebuilding almost the whole edifice of modern astronomy, which cannot be said of any other collection; because to the merit of an exactness which has been seldom attained, and never surpassed, it adds the advantage of a long series of observations.

Maskelyne, who worked hard to make the Royal Observatory a more orderly and accountable institution than it had previously been, would have greatly appreciated this judgement and the credit he was given as being a model director of a modern observatory. Thinking of the context in which Delambre was writing, as director of a scientific sister institution, we might also suspect that this was a none-too-subtle hint to the Académie and French Government that they too should be supporting the regular publication of observations from their own Observatoire. Overall, this assessment of Maskelyne's work had considerable weight in coming from a figure such as Delambre, himself an expert on astronomical tables, observation and calculation. The éloges were not needlessly eulogistic, especially in the case of foreigners, and Delambre was just beginning to gather materials for his six-volume *Histoire de l'astronomie* (1817–27), in which he did not spare his criticism of some astronomers. He wrote there that

'The historian owes nothing to the dead save truth' and his final volume, on the eighteenth century, had Maskelyne's achievement as the conclusion or, possibly, culmination of this long historical study.

Delambre, an astronomer but a foreigner not directly involved in British disputes, also provided an interesting assessment of the conflict with Harrison. He understood that Maskelyne and the Board of Longitude had to make a financial investment that would support the 'two useful methods' they had trialled, recognizing that they were complementary, or 'calculated to assist each other', rather than rivals. However, he considered lunars the more reliable method over long distances and believed that Harrison had been fortunate in getting, eventually, money up to the amount of the full £20,000 reward offered under the Longitude Act of 1714: moreover, that this was evidence of official generosity rather than justice being done. 'It is possible', he added, 'that in this dispute between mechanics and astronomy both sides went a little too far' but he exonerated Maskelyne, who, in 1765, 'appears to us to have displayed as much justice as discernment in assigning one half of the reward to Harrison for his timepiece, and the other half to the lunar tables'.

Delambre's éloge concludes, 'Thus we have described the philosopher: but the man, the father, the friend, was not less valuable. Every astronomer, every philosopher, found in him a brother.' Like Kelly, he gives an impression of Maskelyne as someone who was always ready to help where he could, above all through practical advice and support within his extensive circle of correspondence, and his personal and scientific networks: 'Of a character friendly and amiable, he gained the affections of all those who had the good fortune to know him.'

Here and in his *Histoire*, Delambre acknowledged the assistance of Margaret Maskelyne in supplying information about her father; presumably a version of the 'Autobiographical Notes' but perhaps with additional and more personal information. Correspondence between Margaret and her father's friend, Samuel Vince, reveals that Joseph Banks (1742–1820), as President of the Royal Society, was responsible for sending information about Maskelyne to Delambre. Banks had requested materials from members of the Board of Longitude, who would have known him well, and Vince wrote to Margaret, knowing that she possessed an account. He had himself initially intended to publish this along with some of Maskelyne's unfinished manuscripts.[12] Vince later suggested that this information also be sent to Kelly, and there is thus a family likeness, as well as differences, between these various accounts.

Samuel Vince had made suggestions about what Margaret Maskelyne should include in the account of her father that would be sent to Delambre, principally:

> An Account of such private Circumstances as have not yet come to the Knowledge of the public; such as, of whom and where he was born; where he was Educated, when and to what College in Cambridge he went; after his Degree, where did he go to reside; what was his Employment; when he was elected fellow of Trinity; in short the principal Circumstances of his Life, till he was elected Astronomer Royal.

Delambre, Vince said, knew Maskelyne's public scientific life well enough to 'fill up the rest', although he also suggested that she mention 'his various correspondents [*sic*] with the most distinguished men of his Time, with an Acc^{t}. of the Various Medals he had received for his Public Services'.[13]

Vince seems to have decided that he lacked the time to write a biography himself, or that Delambre's would serve the purpose, although he urged Margaret to continue adding to the manuscript account and to list all her father's publications. Vince also took it upon himself to add that she should 'let [Delambre] know that Dr. M. was not like the French Philosophers (they may take it as they please) who whilst they observed God in his works denied his Existence. I thought this proper'.[14] While Vince's suspicions of the possibly atheistic and revolutionary French *philosophes* might be expected, it is interesting that, in the event, Maskelyne's daughter seems not to have made reference to his religious faith or clerical position.

MASKELYNE'S PLACE IN THE HISTORY OF ASTRONOMY

Vince had promised to prepare Maskelyne's scientific papers for the press, out of 'the Respect I feel for my old, my worthy and greatly esteemed friend'.[15] However, none were to appear and in 1832 a letter to Margaret from the astronomer and historian of astronomy, Stephen Rigaud, expressed a fear that Vince had actually destroyed 'the greater part of your father's papers'. Rigaud warned Margaret to 'collect and preserve all the particular notices which you can (before it is too late) of so distinguished a character as Dr. Maskelyne'. A 'history of his life' would be 'most important as connected with the progress of astronomy for more than half a century'. This she seems to have done, as the surviving Maskelyne material testifies, although what she gathered was rather more personal than scientific. Nevertheless, Maskelyne's published works were ample for him to achieve a position in histories of astronomy. Rigaud had feared that 'knowledge of him as a man' might be lost, but noted that 'Dr. Maskelyne's name can never, however, be lost as an astronomer'.[16]

As we have seen, Maskelyne appeared in the magisterial history produced by Delambre. He was also included, in 1782, in Jean Sylvain Bailly's *Histoire de l'astronomie moderne*, for his work at Schiehallion in 1774 on the gravitational attraction of mountains. Bailly's direct praise of this project in a letter to Maskelyne in 1775 was also high.[17] Later comments by Maskelyne on Bailly's work – this time his *Histoire de l'astronomie ancienne* (1775) – appear to have been less complimentary. According to a book on Hindu astronomy by John Bentley, published in 1825, Maskelyne had written in a private letter that 'Bailly was a pleasing historical writer; but he had more imagination than judgement, and I know that he was condemned by his friends La Lande and La Place, as a *superficial astronomer*, and a very *indifferent calculator*'. This opinion is taken as good by Bentley, for 'It is well known that Dr. Maskelyne was an astronomer of the first-rate abilities, and of the utmost integrity'.[18]

By and large Maskelyne featured in later histories as part of the lineage of astronomers who had made Greenwich, and Britain, significant places for practical astronomy. His combined work on instruments, observation and calculation techniques were key features of this tradition. In addition, he frequently received credit for the changes he brought to the Royal Observatory: establishing routine, ensuring focus on the utilitarian purposes for which it was founded and, above all, guaranteeing regular publication.

An example is in the 1851 *History of Physical Astronomy* by Robert Grant, where he considered the relationship between practice and theory, which had become a key concern for British astronomers in the first half of the nineteenth century as they sought to catch up with Continental theoretical advances. Grant, however, aimed to state the importance of both traditions and their mutual dependence:

> The results achieved by the critical acumen of Halley, the sagacity of Bradley, and the unrivalled practical skill of Maskelyne, advanced at an equal pace with the analytical researches of the geometers on the Continent, and are imperishably associated with them in the magnificent triumphs which illustrate the history of physical astronomy during this period.

Maskelyne's *Greenwich Observations* are, in the tradition of Delambre, singled out for 'their uniformity, their continuity, and their general accuracy', which made them 'of inestimable value in many important investigations relating to astronomical science'.[19]

Seeing Maskelyne as part of this practical rather than theoretical tradition of astronomy in some ways challenges the kind of view of him that is presented in Sobel's *Longitude*. While he is presented there as an academic mathematician and astronomer, Maskelyne and the British astronomical tradition were celebrated for their practical, basic and even humble work. This seems to make him less different from the artisan Harrison than we might expect, especially given the popular influence of Sobel's account, relying as it does on casting him as the clockmaker's 'academic' bugbear. In certain contexts, for example works aimed at educating and inspiring a lower-class reading public, such as the *Gallery of Portraits* published in the 1830s by the Society for the Diffusion of Useful Knowledge, Maskelyne's hard work and dedicated routine, undertaken for public utility, were greatly celebrated.[20] In this biographical collection, he is presented as a champion of work that is useful and practical, and which considers the ordinary sailor as well as benefiting commerce and the prosperity of the nation. At a similar period, because of his role in disputes with the autocratic President of the Royal Society, Joseph Banks, Maskelyne could be championed by reformers as standing for real science and professional men against the dilettanti and wealthy elements of the Society that Banks had represented.

Given the accepted importance of the Greenwich Observatory in the nineteenth century, and the emphasis on routine and utility that was championed by George Airy as Astronomer Royal from 1835 to 1881, it is not surprising that Maskelyne maintained a steady position within general histories of astronomy. In such accounts his legacy was focused on positional astronomy and, while a nod may have been given to its importance for practical navigation, there was no need for it to be complicated by detailed consideration of longitude hopefuls. His entry in the original *Dictionary of National Biography* (1893), written by the historian of astronomy, Agnes Mary Clerke, might stand as the last word for the century in which he died. The outline is familiar, his disputes are minimized – 'The discharge of his onerous task of testing timepieces exposed him to unfair attacks, especially from Mudge and Harrison, against which he defended himself with dignity' – and he is described as indefatigable and 'of a mild and genial temper and estimable character'.[21]

HARRISON'S REPUTATION

To understand the change in Maskelyne's reputation beyond such biographies and disciplinary histories, it is necessary to look at what happened to Harrison's. Famous in his lifetime, he too retained a presence in biographical dictionaries and encyclopaedias after his death. The narrative of provincial, lower-class boy made good (not to mention extremely wealthy) was appealing, especially in publications aimed at an aspiring working class. He provided an excellent exemplar for the workers who were told to better themselves by gaining scientific education, by reading or attending lectures, and applying that knowledge to the work of which they already had intimate and practical experience. Harrison's 'genius', 'natural abilities' and dedication overcame the 'impediments' of birth.

Some inspiring rhetoric made its way into the *European Magazine*'s biography of Harrison, published with an accompanying portrait in 1789 (Fig. 3). His life was proof that 'poverty may learn that the efforts of genius will ultimately prevail over every difficulty, genius may be taught industry, and industry encouraged to perseverance'. The difficulties he faced were not ascribed to the Board of Longitude, but rather to circumstances – the fact that 'he was no man of the world' and 'no writer'.[22]

This account provided a basis for another, published in the *General Biography* edited by John Aikin, in 1804.[23] This, however, made some interesting additions, including the point that, although he had to be persistent, the Board of Longitude had ultimately given him more than £20,000 in total, and that he had also received money from the East India Company. This biography explained that the 'delay in issuing to him the full reward, originated in the anxiety of the commissioners of the longitude to do justice to the public, at the same time that they encouraged merit in an individual'. The author understood that their concern was for Harrison's watch to be 'rendered ... of general use' through being produced by other makers and at a much cheaper price. Maskelyne is not personally mentioned, and the Board is not generally criticized, but it is worth remembering that at this date he was still alive and the Board was still functioning.

Well after the deaths of both Harrison and Maskelyne came the publication of a Harrison biography in the SDUK's *Gallery of Portraits* in 1835.[24] Not surprisingly, given its intended audience, this carried the 'self-help' story of the provincial artisan, who 'is, on the whole, a fine instance of the union of originality with perseverance'. SDUK biographies always tended to focus on the need for hard work, as well as natural talent – perspiration rather

Fig. 3: This engraving by P.J. Tassaert was based on the 1767 oil portrait of John Harrison by Thomas King. It was published, with a biography, in the Society for the Diffusion of Useful Knowledge's *Gallery of Portraits* in the 1830s.

than inspiration. However, it was also typical of the Society's output that biographies should admit to, and strongly censure, bad behaviour. In this case, Harrison's 'absurd conduct' toward the Board of Longitude was highlighted and he was criticized for wanting things both ways: 'Harrison desired, in addition to the large reward claimed by him, to have a monopoly of the manufacture of his watches.'

Alongside such accounts, there is evidence for interest in Harrison from those who would have seen themselves as working in his footsteps, although much of this was probably in oral tradition or specialist literature rather than aimed at broader audiences. Exceptions include the curious book by John Harrison's grandson of the same name, who wrote under the pseudonymous anagram 'Johan Horrins'. This was called *Memoirs of a Trait in the Character of George III* (1835) and told a partial and unreliable version of the story of Harrison's relationship with Maskelyne and the Board, leading to the king's involvement in his case. This continues the depiction of Maskelyne presented in the Harrison 'Journal', quoting extensively from the earlier polemical writings, asserting his bias, 'double-dealing' and pecuniary interest in the 'rival' method of lunars.[25]

A more straightforward brief account, in which the Horrins book was a reference, appeared in Edward J. Wood's *Curiosities of Clocks and Watches: From the Earliest Times* (1866).[26] This described events without casting clear judgement against either side, merely saying that £20,000 was, eventually, the just reward for genius and perseverance and that Harrison was ill-equipped to explain his ideas. With just a hint a censure, it claimed that Maskelyne had insisted on his right to be present at Harrison's explanation of the mechanism of his sea watch (H4), required by the Board of Longitude in 1765.

Harrison's special place in the history of British horology was guaranteed, just like Maskelyne's in the history of British astronomy. We see it in action when the clockmakers, Dent & Co., agreed in the 1830s to clean and repair the neglected Harrison clocks without payment, in honour of the objects and their creator (although this was, admittedly, only after they had for some years ignored the Astronomer Royal's instruction to do so). The heritage of horology received a significant boost from the opening of the object collection of the Worshipful Company of Clockmakers to the public in 1874. Harrison was always to be a key figure there: the Company acquired his regulator in 1877 and his H5 watch in 1891. It was also responsible for reconstructing his tomb in Hampstead in 1879, repairing it in the 1930s and, much more recently, lobbying for and funding a memorial in Westminster Abbey.

THE HISTORY OF HOROLOGY

The full rediscovery of Harrison's legacy was largely down to Rupert Gould, who located, cleaned, restored and championed the Harrison timekeepers, which had again been left languishing in poor conditions at the Royal Observatory since their minor overhaul by Dent's in the nineteenth century. This work was undertaken as he was researching his book, *The Marine Chronometer, its History and Development*, published in 1923. Gould was a Royal Navy man, having been a navigating officer and later working in the Hydrographic Service: he thus represents another constituency within which interest in antiquarian horology, and Harrison, developed. Gould began his specialist interest in this area as one of a range of other technical, scientific and 'sceptical' topics.

Given that Gould approached the story from the direction of interest in the development of chronometers, it is unsurprising that his account of the disputes with the Board of Longitude and Maskelyne gave the Harrisons' version of events a stronger hearing than any of the texts discussed so far. He did, however, acknowledge that their version was not entirely justified. Nevertheless, taking the story from Harrison's point of view, Maskelyne appeared as the 'declared and bitter enemy' of the Harrisons. Gould believed that Maskelyne, 'even if unbiased, was far too harsh' in the conduct of the trials of Harrison's timekeeper and in his demands, and that he would have been happy to see the clocks fail. Such a judgement is not borne out by the archival evidence or, indeed, by the fact that lunars and timekeepers were recognized as complementary methods.

Gould not only wrote an important and authoritative book on the history of the chronometer; he also seems to have been very good at inspiring those he conversed with or lectured to regarding Harrison and his timekeepers. The restored clocks, too – the most charismatic of objects – played an important role. A good example is a lecture given by Gould in 1935, at which he placed all five of Harrison's marine timekeepers and Kendall's (K1) copy of H4 before him, all alive and ticking. Gould played a key role in getting the clocks (which then belonged to the Admiralty and now to the Ministry of Defence) on display in the National Maritime Museum from its public opening in 1937. This lecture was part of his campaign to do so and it had an interesting effect on the Astronomer Royal of the time, Harold Spencer Jones, who afterwards brought up 'the question of Maskelyne's alleged "conflicts of interest" in his dealings with Harrison', and apparently 'professed himself anxious to do penance for the misdemeanour of his predecessor'.[27] It was perhaps this that later led him to do some investigation into how Maskelyne had treated H4 during its trial at the Observatory. He felt that this exonerated Maskelyne from accusations of serious foul play.[28]

The first full-length biography of John Harrison, subtitled *The Man who Found the Longitude*, was published in 1966 by Humphrey Quill, a retired colonel in the Royal Marines and Senior Warden of the Clockmakers' Company. He wrote it after deciding 'that of all the great men who were the pioneers of precision timekeeping, the name of John Harrison stood head and shoulders above all others of his profession'. Despite this enormous admiration, his reading of Harrison and his context led him to a generous understanding of Maskelyne's and the Board's decisions, given the risk of gambling a large amount of public money on a unique object. He was clear, too, that Harrison and his 'sulky non-cooperation' were difficult to deal with. He concluded that 'John Harrison and his son seem to have persuaded themselves, without apparent justification, that they were being victimized by the Board.'[29]

Quill's work has rightly been described by the historian of science, Jim Bennett, as 'a model of balanced judgement and sympathetic appreciation of the different situations of the historical characters'.[30] Nevertheless, reading, or at least hearing about, Harrison's story still led to the idea of his having been victimized. In the book's introduction, for example, the new Astronomer Royal, Richard Woolley, described it as 'a story of a somewhat discreditable kind', as far as the Board of Longitude was concerned. This is certainly not Quill's conclusion and it suggests the sticking-power of the Harrison version of events, once heard, especially in an age that prefers to champion merit over privilege. There again seems to be an inherent tension between what the documentary evidence shows the story to be and the way it gets told or displayed in popular or informal contexts. The appeal of the clocks, which are visually much more arresting than astronomical tables, have long given Harrison's version more currency than Maskelyne's.

Both attitudes toward Maskelyne seem to have co-existed within the National Maritime Museum after it became responsible for Harrison's clocks and, later, the Royal Observatory itself. While one former curator, Derek Howse, eventually produced a full-length and sympathetic biography of him, and there was an exhibition to celebrate the bicentenary of the *Nautical Almanac* in 1967, there was also plenty of interest in gaining better recognition for Harrison.[31] The latter impulse led to a 1962 exhibition, 'Four Steps to Longitude: Harrison', which focused on the four marine timekeepers, relegating lunar distances to a separate section on 'The rival method'. There was also a small exhibition in 1976, marking the bicentenary of Harrison's death. One of those involved with this display, Will Andrewes, was also motivated to mark the tercentenary of Harrison's birth in 1993. By then curator of the Collection of Historical Scientific Instruments at Harvard, he convened a symposium on 'the history of finding longitude at sea'. This gathering led to two publications, one a book of the conference papers, *The Quest for Longitude* (1996), and the other Sobel's *Longitude*.

Sobel had attended the conference not as a historian or curator but as a science journalist. She had been sent there by the publishers of the *Harvard Magazine*, in which her initial account was published in 1994. The potential of the story was recognized by a publisher working for Walker and Company, who contacted Sobel to suggest she turned it into a book. The story, known to specialists from one side or another, was thus repackaged as a coherent whole for a popular audience for whom it was entirely new. In the practical outcome, the honours were also fairly even between her skill in turning it into a relatively brief and appealing account, and the innovative 'pocket-book' format in which it was presented and marketed by her publishers.

If we consider the symposium as Sobel's source material, it also becomes clear that her heroic Harrison and villainous Maskelyne were less a convenient approach for a popular narrative and more a fairly accurate representation of what many of the experts present had indicated. In *The Quest for Longitude* the astronomy story is largely, and inaccurately, relegated to a section on 'Early attempts to find Longitude'. The bulk of the book is focused on the horological side, several chapters being the work of horologists with great technical knowledge of their subject but perhaps less interest in the wider context. Maskelyne gets much less space than one would expect, either in relation to astronomy or in his role as a Commissioner of Longitude, in which he actually gave considerable support to the timekeeper method, even as he developed a poor relationship with some of the clockmakers. By and large he appears simply as Harrison's *bête noir* and, following Harrison, as the key enemy within the general foe of the Board of Longitude. Phrases like 'insult to injury', 'aided and abetted', 'harsh and bigoted' abound.

Jim Bennett's review of *The Quest*, for the journal *Antiquarian Horology*, included considerable criticism of it as a work of historical scholarship. He found much of the volume to be 'speculative apology for Harrison', with the man apparently 'becoming a "cause" rather than an object of historical study', which led to a 'one-sidedness' in analysis of Harrison's disputes with the Board of Longitude. Its 'worrying enthusiasm' was, indeed, a regression from the work of Gould and Quill.[32]

Bennett received some touchy responses to this review, which were published in subsequent issues.[33] Several of these correspondents evidently identified strongly with Harrison. This meant that, on the one hand, they felt that their hands-on experience and practice were more important than a reading of historical texts and contexts and, on the other, that their being criticized by an academic was an echo of Maskelyne's supposed dismissive treatment of Harrison. Defending the critics rather than champions of an underdog, or the judgement of more established people, institutions and disciplines, can easily lead to charges of elitism. While most historians and many horologists would today agree that Maskelyne was unfairly maligned in Sobel's account and parts of *The Quest*, in non-expert contexts this view nevertheless remains current and colours discussions of the history of longitude and the Royal Observatory, and of Maskelyne himself.

Because of its huge success, historians of science have felt a need to respond to Sobel's book.[34] The main issue relates to the fact that a simplified story of a lone genius, of one man working against the establishment, can and nearly always does sweep aside the more nuanced accounts of combined effort that professional historians of science prefer to tell. David Miller suggested that such narratives respond to 'the ideological needs of scientific culture', despite seriously misrepresenting how science actually works, through collaboration, testing and consensus.[35] The casualty is the hard work of historians of science, who find themselves wondering if their preferred narratives can ever be similarly popular. Maskelyne was associated with work that was supremely collaborative, crossing centuries and nations as well as drawing on the efforts of other individuals, and his most significant abilities were in management, organization and communication. These are necessary factors in understanding the way that science works, but can be difficult material from which to form a compelling narrative.

WHERE NEXT?

There are, today, some signs that there will be a further swing in perceptions of Maskelyne. As regards popular history, the time appears to be ripe for a rescue of his reputation, partly because Harrison's has ridden so high. For readers who need a new story, then a more sympathetic portrait of Maskelyne and account of the lunar-distance method are worth a hearing. Within the National Maritime Museum there has been opportunity to think more about him, with the arrival of an important collection of family manuscripts and artefacts, and a renewed internal interest in telling the story of the Royal Observatory, alongside related exhibition and academic projects exploring the history of the Board of Longitude. This volume is one outcome of this convergence of factors.

Harrison remains an important part of all these stories but recent work has underlined Maskelyne's equal or greater importance in the story of longitude. It has also brought more attention to his relationship with not just one clockmaker but also with colleagues, assistants and family, as well as the full range of his work and legacy. Research has brought out his contributions to the development of instruments, techniques, training and observational regimes, both in permanent observatories and on scientific expeditions. It has become clear how important he was in the development of the British tradition of exploratory voyaging, ensuring a wide range of scientific work was undertaken, as well as in aiding navigation and survey by use of the sextant, *Nautical Almanac* and timekeepers.

One of the obvious conclusions has been that the version of Maskelyne presented in *The Quest* and *Longitude* is very wide of the mark. The archives held in the NMM and Cambridge University Library have confirmed Howse's portrait of a man who was, by and large, fair, ready to help and conscious of obligations. He was a doting father; he aided his assistants and computers often above and beyond what we might expect; we find him privately described as pleasant and jocular, and see him doing his best to fulfil his obligations even when placed in a difficult situation. Admittedly, scholarship also reveals that biographies are written to suit particular times, people and purposes. The early biographical accounts of Maskelyne are generous, as we would largely expect for the period, especially of someone alive or only recently dead. Nevertheless, in assessing the Harrison disputes, the relationship between clocks and lunars, and in giving a sense of the importance of Maskelyne's work and portraying the man, these accounts are not so far removed from the well-researched biographies by Howse and Quill.

Maskelyne's popular reputation today is, in large part, the casualty of a good story, underlain by a developing interest in, and desire to promote the reputation of, John Harrison. However, it is also influenced by the fact that astronomical navigation has been largely superseded and that the positional astronomy of Greenwich is not the kind that excites today's audiences. The attempt to humanize Maskelyne, to understand the range of activities in which he was involved and to appreciate his managerial abilities would seem to be an appropriate response.

FURTHER READING

All of the eighteenth- and nineteenth-century biographies discussed here are available online through Google Books. In addition to the various twentieth-century biographies of Maskelyne, Harrison and Gould, the accounts of the longitude story and the critical discussion of *Quest* and Sobel cited in the endnotes, a useful essay is J. A. Bennett, 'Science lost and longitude found: the tercentenary of John Harrison', *Journal of the History of Astronomy* 24 (1993), pp. 281–87. Information about exhibitions held at the National Maritime Museum can be found in Kevin Littlewood and Beverley Butler (eds), *Of Ships and Stars: Maritime Heritage and the Founding of the National Maritime Museum* (New Brunswick: The Athlone Press, 1998).

For the history of scientific biography and their use in understanding the changing reputations of scientific figures, see *Telling Lives in Science: Essays on Scientific Biography Telling*

Lives in Science edited by Michael Shortland and Richard Yeo (Cambridge: Cambridge University Press, 1996); Thomas Soderqvist (ed.), *The History and Poetics of Scientific Biography* (Aldershot: Ashgate, 2007) and Rebekah Higgitt, *Recreating Newton: Newtonian Biography and the Making of Nineteenth-Century History of Science* (London: Pickering and Chatto, 2007). Looking from the perspective of science communication, see also Davida Charney, 'Lone geniuses in popular science: the devaluation of scientific consensus', *Written Communication*, 20 (2003), pp. 215–41.

There is a useful study of the Académie des Science's tradition of éloges in Charles B. Paul, *Science and Immortality: The Éloges of the Paris Academy of Sciences (1699–1791)* (Berkeley, Los Angeles & London: University of California Press, 1981). The role of histories in the formation or consolidation of scientific disciplines has been discussed by Rachel Laudan in 'Histories of the sciences and their uses: a review to 1913', *History of Science*, 31 (1993), pp. 1–34.

CASE STUDY A

THE LONGITUDE PROBLEM

As will already be amply clear from the introduction and first chapter, much of Maskelyne's working life revolved around the question of how to establish longitude. In order to make sense of his career, it is therefore useful to understand the nature of the problem and of the potential solutions to it with which he engaged. This was, and remained, a core business of astronomers, especially those funded by the state and focusing on issues of presumed utility, such as navigation and cartography.

Longitude is one of the two co-ordinates needed to fix any global position. Lines of latitude are circles around the Earth, parallel to the equator, while lines of longitude run from pole to pole. Each can be expressed as a number of degrees (which are subdivided into minutes and seconds) away from a given reference point, either north or south from it (latitude) or east or west of it (longitude). Latitude has a natural reference point in the equator but longitude does not (Fig. 1).

Only agreed convention defines a particular starting point for measuring longitude. Often Cape Verde was used, as a division between the eastern and western hemispheres, or old and new worlds, but capital cities or ports of departure were also common reference points. That the world's prime meridian today runs through Greenwich is, in large part, a result of the work that Maskelyne did.

Because of its natural reference point at the equator, and the regular relationship between that and the Sun or stars, latitude is relatively easy to find by astronomical observation. It can, for example, be found with reasonable accuracy by observing the angular height of the Sun at midday: at the equator it is almost directly overhead but appears at declining angles the further north or south you go. By taking account of the date and the predicted position of the Sun for that day, it is possible to calculate your latitude.

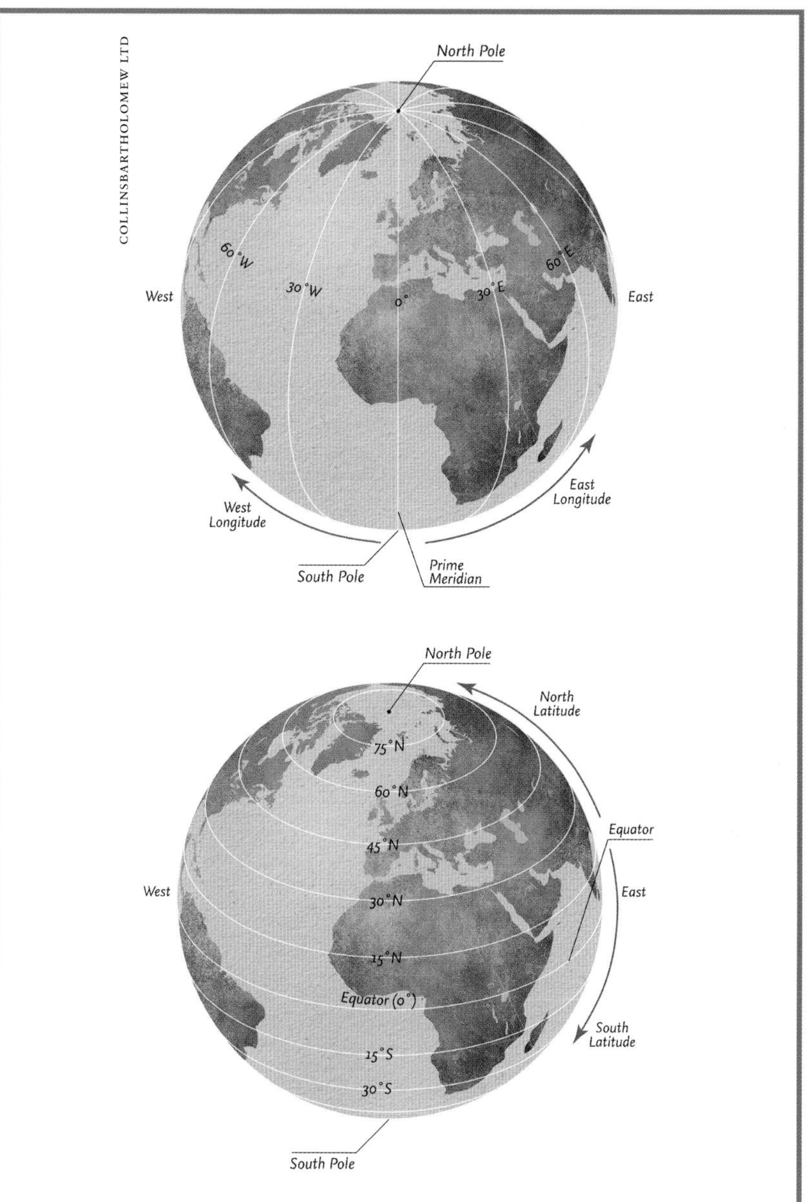

Fig. 1: Longitude and latitude lines form an imaginary grid on the Earth's surface. Lines of longitude run north-south, from pole to pole, and give the distance east or west of a zero line, known as the prime meridian. Latitude lines run around the globe parallel to the equator and give the distance north or south of that point.

In the absence of such natural references, establishing longitude is considerably more difficult. However, even long before mechanical clocks existed it was understood that, as a result of the Earth's daily rotation, the difference in longitude between two points is equal to the difference between their local times. If the Earth rotates 360° of longitude in twenty-four hours, then each hour of difference in local time equals 15°, and 1° of longitude is equal to four minutes' difference in local time (Fig. 2).

It is also possible to determine local time by measuring the angle of the Sun or stars. The question was how to compare this local time with that in another, known location. This is easy to do today, in an era of instant communication and accurate portable timekeepers, but it was a technical problem that exercised many minds and hands for centuries. It was made even more difficult by conditions at sea: motion and changing temperatures

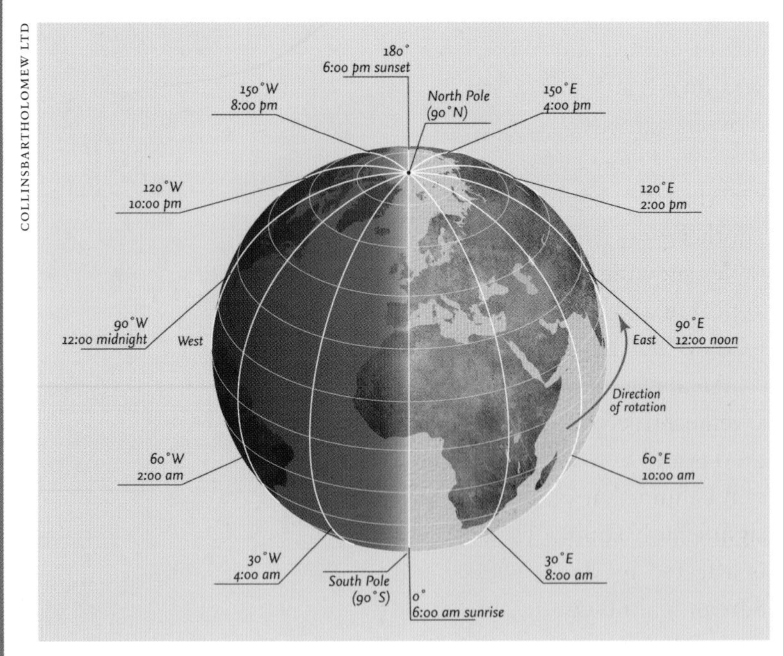

Fig. 2: The difference in longitude between two places is equivalent to the difference in local time. Each 15° of longitude is equal to an hour's difference in local time. A local time ahead of that at your position would indicate a location to the east, and, if behind, one to the west.

during voyages were particular challenges to the use of precision instruments.

It was obvious that having a clock that could keep time accurately, for a sufficient period and in all conditions, would be a solution. However, it was not until the seventeenth century that even land-based clocks were accurate enough, and not until the mid-eighteenth that techniques and innovations were developed that could create marine timekeepers. Only in the nineteenth century were these timekeepers, by then known as chronometers, sufficiently numerous and robust to be of widespread use.

In the meantime, astronomy provided the most likely solution. Certain celestial events, such as eclipses, are visible at the same moment from different parts of the Earth: observing these, then comparing the local times at which the event occurred, provides a measure of the difference in longitude between the observing locations. Such events happened too irregularly to aid navigation but from the sixteenth century onward there was investigation into whether the position of the Moon, as it moved against the background of the stars, could be predicted accurately enough to be used as a celestial timekeeper. The idea was that the angular distance between the Moon and the Sun or another star, could be measured and compared with predictive tables showing at what local time this same measurement would appear at a known location (Fig. 3). This was the 'lunar-distance method' but its complexities were not to be tamed for centuries.

An alternative celestial timekeeper was discovered in the early seventeenth century, when Galileo first observed four of Jupiter's moons and noted their regular eclipses as they moved in front of and behind that planet. The motions of the Jovian moons were significantly simpler to predict than that of Earth's Moon, which is affected by the gravitational pull of both the Earth and the Sun and, although it took the work of several astronomers to do so, by the end of the seventeenth century these were sufficiently well understood for them to be used to establish longitudes on land. They were not used at sea because the angles involved were very small and impossible to observe at sufficient magnification

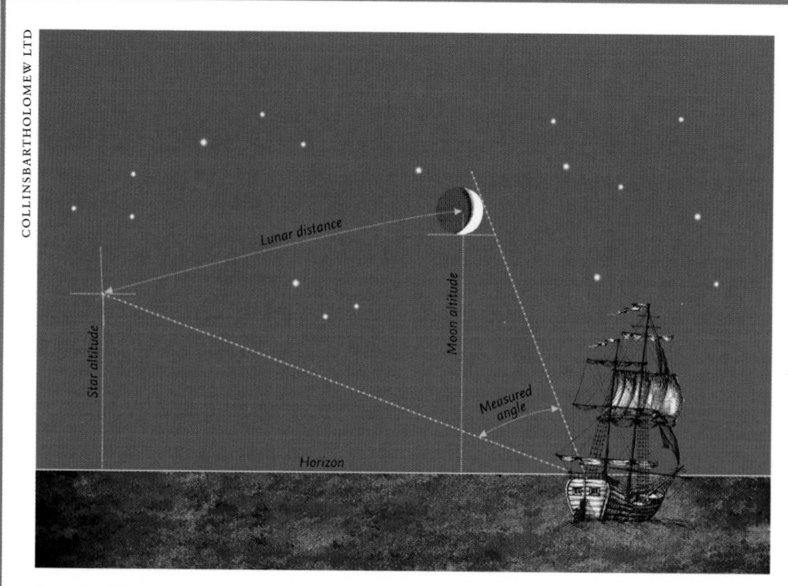

Fig. 3: The lunar distance measurement is the angular distance between the Moon's centre and a star or the Sun, which varies over time. The altitude of both objects must also be measured, if possible at the same time. These measurements then had to be processed by taking account of local time, and the effects of refraction and parallax before they could be compared to predictions of the Moon's position as observed at a reference location.

from the moving deck of a ship, although many people tried to come up with chairs, mechanisms and platforms to steady the observer and his telescope.

Because observing the much larger angles between the Moon and Sun or other stars was considerably simpler, interest continued in the lunar-distance method. To become feasible, better maps of the stars were required and understanding of the motion of the Moon, or its theory, had to be vastly improved. The need to do this work was the reason that official observatories were founded in Paris (1666) and at Greenwich (1675). It was also one of the reasons that Isaac Newton undertook theoretical work on the motion of the Solar System, and his inverse square law of universal gravitation was a vital step in the process. Nevertheless, the complexities of the Moon's motion stumped even Newton.

This job was left for the generations that followed, requiring both more observational work and the development and application of new mathematical techniques. This coming together of observation and theory, to create a predictive theory of the Moon's motion that was, if not perfect, at least good enough, took place just as Nevil Maskelyne left Cambridge to embark on his career in astronomy. In 1757, Tobias Mayer (1723–62), a professor at Göttingen University, had sent his tables of the Moon to Admiral Lord Anson, who was then First Lord of the Admiralty. Mayer explained that the tables were good enough to predict the Moon's position with accuracy sufficient for navigational purposes, provided that several observations were made and subjected to a lengthy series of mathematical corrections and calculations.

Mayer contacted Anson, through an intermediary, because he was *ex-officio* one of the Commissioners of Longitude, who had originally been appointed by the British Government by Act of Parliament in 1714 (Fig. 4). Anson naturally referred the tables to the chief astronomical expert among the Commissioners, the Astronomer Royal, James Bradley. After comparing Mayer's tables with his own observations at Greenwich, Bradley was impressed. As a result, the tables were soon being tested at sea, in 1757 and 1759, by one of Anson's former officers, Captain John Campbell. These tests were not full trials, as Campbell undertook them while on active blockade duty, but his work using two still-new observational instruments – a Hadley's quadrant (or octant) invented in the 1730s and Mayer's own repeating circle – prompted him to suggest improvements that led to the development of the sextant (Fig. 5).

The Commissioners of Longitude had been appointed to adjudicate on ideas and devices submitted for a reward of up to £20,000, depending on how accurately they managed to establish longitude after a sea voyage to the West Indies. Interpreting the 1714 Longitude Act broadly, the Commissioners had already dispensed much smaller amounts of money to encourage or repay expenses of those engaged in what they considered useful

Anno Regni

A N N Æ

R E G I N Æ

Magnæ Britanniæ, Franciæ, & Hiberniæ,

DUODECIMO.

At the Parliament Summoned to be Held at *Weſtminſter*, the Twelfth Day of *November*, *Anno Dom.* 1713. In the Twelfth Year of the Reign of our Sovereign Lady *ANNE*, by the Grace of God, of *Great Britain*, *France*, and *Ireland*, Queen, Defender of the Faith, *&c.*

And by ſeveral Writs of Prorogation Begun and Holden on the Sixteenth Day of *February*, 1713. Being the Firſt Seſſion of this preſent Parliament.

L O N D O N,

Printed by *John Baskett*, Printer to the Queens moſt Excellent Majeſty, And by the Aſſigns of *Thomas Newcomb*, and *Henry Hills*, deceas'd. 1714.

Fig. 4: 'An Act for Providing a Publick Reward for such Person or Persons as shall Discover the Longitude at Sea'. The first Longitude Act received the Royal Assent of Queen Anne on 9 July 1714, only weeks before her death.

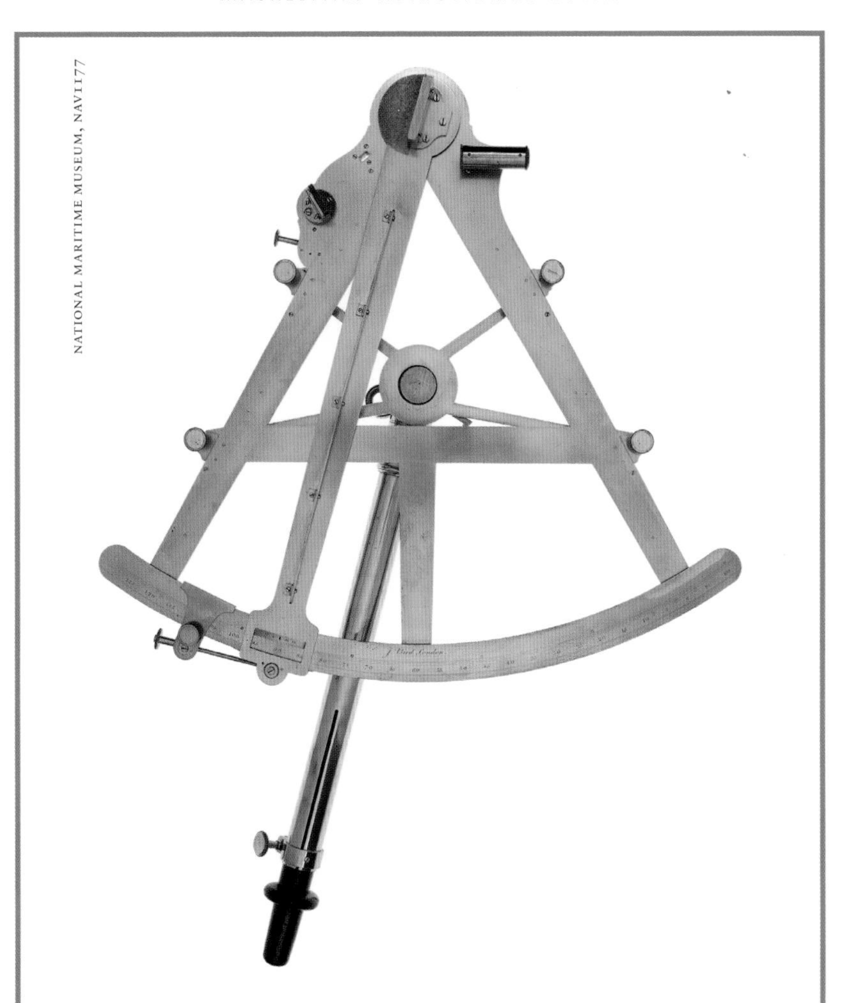

NATIONAL MARITIME MUSEUM, NAVI177

Fig. 5: Sextant by John Bird, c.1758. Bird was commissioned to make such an instrument by the Board of Longitude, following the suggestion of Captain John Campbell.

or interesting work. The most substantial beneficiary so far was John Harrison, a clockmaker who had already had some success with a marine timekeeper, now known as H1 (Fig. 6), which had been tested on a voyage to Lisbon in 1737.

The Commissioners were clearly impressed and were persuaded by Harrison that he could improve on his machine. Between 1737 and 1757, he received £2750 toward the completion, testing and adjustment of his later timekeepers, H2 and H3, neither of which ever went to sea, and his new sea watch, H4.

The time could not have been more propitious for someone with the skills to make the kinds of observations that would test these methods to build a career. This is what Maskelyne was to do, starting in 1761, when he tried Mayer's tables on his voyages to and from the island of St Helena, and continuing in 1763–64, when he took part in the official Board of Longitude trials of the tables, Harrison's H4 and a marine chair, designed by Christopher Irwin to aid the observation of Jupiter's moons on board ship. Ultimately, as Astronomer Royal and a Commissioner of Longitude from 1765 to 1811, Maskelyne's career was based on finding ways to improve and support the two methods that came out of these official trials successfully: Mayer's lunar tables and Harrison's watch.

NATIONAL MARITIME MUSEUM, MINISTRY OF DEFENCE ART COLLECTION, ZAA0034

Fig. 6: John Harrison's first sea timekeeper, later known as H1, was completed in 1735 and tested in 1737. It performed remarkably well, but Harrison abandoned it and convinced the Commissioners of Longitude that he could produce a better design.

2

'THE REV. MR. NEVIL MASKELYNE, F.R.S. AND MYSELF': THE STORY OF ROBERT WADDINGTON

Jim Bennett

INTRODUCING ROBERT WADDINGTON

Nevil Maskelyne's assistant on his expedition to St Helena to observe the transit of Venus of 1761 was Robert Waddington. Appointed by the Royal Society, he was not their first choice for the job but replaced Charles Mason, after Mason had declared himself willing to lead a second expedition, bound for Bencoolen in Sumatra.[1] (On the transit expeditions see Case study B.) At Bencoolen, Mason was to be assisted in turn by the surveyor and astronomer Jeremiah Dixon. Mason had an established position in this circle of activity and expertise, being assistant to the third Astronomer Royal, James Bradley, at Greenwich. His replacement is more of a puzzle but, despite his relative obscurity, Waddington's career deserves attention. It might be said that it deserves attention *because* of his obscurity, since we do, in fact, have a fair amount of documentation, giving us unusually rich access to one example of what might be called a jobbing or opportunist career in mathematics in eighteenth-century London.

By seeking to make a career for himself among the range

of possibilities offered in the world of public and commercial experimental philosophy, and – more especially in his case – mathematical practice, he was far from being alone. There were many who followed similar ambitions. We might hesitate to say that Waddington was typical of the profession only because there were so many different trajectories. What were typical, perhaps, were his versatility and opportunism.

Of the four observers on the pair of expeditions, three are sufficiently well known to feature in the *Oxford Dictionary of National Biography*. Of these, only Maskelyne was a university graduate and there is no evidence that Waddington, who is the one not included, had a university education. The letter he wrote on reaching St Helena to the Secretary of the Royal Society, Thomas Birch, when he was certainly working with care and attention to detail, and would surely have been on his best grammatical behaviour, provides some indication of Waddington's educational background.[2] He begins: 'These comes to inform you of our safe arrival here' and in his account of the voyage he uses such expressions as 'ye decks was immediately clear'd'. Combining a singular verb with a plural subject was common enough in everyday speech but in a relatively formal or semi-official report was (and remains) an educational marker. It is a small pointer but we need to deploy all the evidence we can find about Robert Waddington. Neither birth information nor his age at death in 1779 are yet known and the earliest records of him uncovered so far are his contributions to what we might call the recreational mathematical press. A 'Mr. Rob. Waddington of Hull' answered a mathematical problem in *The Gentleman's Diary, or the Mathematical Repository* in 1758, while in 1761 a question was contributed by 'Mr. Robert Waddington, Mathematical Instrument Maker in Hull'.[3] 'Mr. R. Waddington' sent a question to Benjamin Martin's *General Magazine of Arts and Sciences*, and Martin published the contributor's own solution in January 1759. It concerned the lengths of pendulums beating seconds in different latitudes, a subject that would be relevant to activity on St Helena.[4] A 'Mr. Waddington, of Hull' contributed further questions, series of daily barometric readings taken at Hull, monthly maximum and minimum temperature readings, and eclipse calculations.[5] The identity of this correspondent is confirmed when Martin reports in December 1760 that the Royal Society have appointed 'Mr. R. Waddington, of Hull' to assist Maskelyne on St Helena.'[6] So we can say that, early in his career, Waddington was a (previously unidentified) instrument maker active in Hull: we will learn from a later episode that, for a time, he also worked as clerk to a merchant in Hull.[7]

A valuable and unused source is a series of letters that Waddington wrote to the gentleman astronomer, Nathaniel Pigott, preserved among Pigott's papers at the Royal Astronomical Society.[8] Pigott was born in 1725, so cannot have been much older than Waddington, but at times Waddington seems to adopt the tone of a teacher, sending Pigott long worked examples of astronomical calculations from his observations, or advice on the use of instruments. We discover that Waddington was living in the Pigott household in Whitton, Middlesex, immediately prior to his appointment by the Royal Society to accompany Maskelyne. A notebook belonging to Birch also has a note in Maskelyne's hand of Waddington's new address.[9] This is in line with an eclipse observation of 22 November 1760 by Waddington, sent to Martin's *General Magazine*, and not made at Hull but 'at Whitton, near Twickenham, in the County of Middlesex.'[10] Howse suggests that Dixon's recommendation to the Royal Society came from the instrument maker John Bird (1709–76) and the same could be true of Waddington, since Bird had supplied Pigott with a quadrant. Waddington proved to be a suitable choice, as we might expect if the recommendation had come from someone as practical, reliable and well connected as Bird.

THE EXPEDITION TO ST HELENA

The earliest surviving letter from Waddington to Pigott was written from St Helena on 26 May 1761.[11] After sending good wishes to all, Waddington added, 'I hope you have made a Considerable progress in Astronomy since my Departure from

Whitton.' He then sent the times of a lunar eclipse observation on the 18th, with a three-foot reflector, which he hoped Pigott had observed as well. However, the letter to Birch was earlier, written on 13 May, and is the principal source for Waddington's experience of his and Maskelyne's voyage from Portsmouth to Jamestown, St Helena.

Waddington left London on 9 January and was on board an East India Company ship, the *Prince Henry*, at Portsmouth on the 10th. The commander, Captain Charles Haggis, took no great notice of Waddington joining his company (other than including him as a passenger in the list of those on board, which was standard practice), whereas when Maskelyne came aboard on the 12th, with the most senior of the three Company 'supercargoes' travelling to China, they were saluted with nine guns.[12] The Seven Years War still being in progress, they set sail on the 17th in a convoy of about one hundred vessels mostly bound for the West Indies, left the Lizard behind on the 21st and parted from the rest of the fleet on 1 February.[13] Waddington was keeping an account of the ship's position and located the leave-taking as in latitude 33° 8' N, longitude 21° W. They were now accompanied only by another East Indiaman, the *Hawke*, under the command of Captain George Kent. Waddington was alerted to the dangers of such slight company only the following morning, when at 9 a.m. they sighted a ship, which made towards them. By 4 p.m. they concluded they were under enemy threat and, with the order 'down Chests & up Hammocks', they cleared the decks and prepared for an attack. In the event the ship, which Kent noted in his journal as 'a Privateer of about 20 Guns', passed about three leagues to windward.[14]

Waddington reported no further excitement until they crossed the equator on 21 February. He was intrigued to find that they were on the equator at noon – 'a thing rarely known to be under ye Line at Noon' – and he put their longitude at 19° W. On 4 March they saw another sail and 'Chests & Hammocks was transpos'd as before', but what they took to be a peaceful Portuguese vessel bound for Brazil passed uneventfully southwards of their position.[15] Within the routine of shipboard life,

Maskelyne occasionally contributed a sermon to Sunday prayers for the ship's company.[16]

Waddington recorded parting company with the *Hawke* on 11 March in 27.5° S, 26° W and we learn from Kent's record that there was an exchange of three cheers between the crews.[17] On 5 April there was another anxious moment, after a sail was sighted from the main masthead at 5 a.m. and, as it headed directly towards them, 'ye decks was immediately clear'd, down Chests & up Hammocks &c'. Over five hours later and after attempts to signal, this turned out to be an East India snow-brig, as Waddington records, whose captain told them that all was well on St Helena, which they sighted the following day and where they disembarked on 7 April (Fig. 1).[18]

Fig. 1: The island of St Helena was a watering place for homeward-bound ships of the East India Company. This painting shows the East Indiaman *Northumberland* from two angles, at anchor near St Helena. It was painted by Thomas Luny in the late eighteenth century.

Waddington gave Birch a detailed account of the progress made during the voyage, first calculated using dead reckoning, with longitude determined 'according to ye Account I kept of ye Ship's Way'. He reports his calculated positions after every fourteen-day interval, giving the latitude and longitude and the distance covered in each section, and concludes that the total voyage was 7,760 miles. He says that the captain complimented him on keeping the best account in the ship and, although by the end he was in error by seven degrees, the accounts of the captain and first mate were even further from the accepted longitude of St Helena.

Waddington then gave Birch a second set of positions, with the longitudes determined by lunar distances. For each of eleven observations between 28 February and 30 March he gave the difference between the astronomical determination and the longitude by dead reckoning, or 'by account': by the end of this series, the lunar determination was 5° 52' W of dead reckoning and, since the latter was seven degrees east of the accepted figure eight days later, Waddington felt justified in concluding the success of his lunar work.

Maskelyne too was keeping a journal involving positions by account and by observation.[19] His first clear record of a lunar fix was for 10 February, when he took an afternoon measurement of the distance between the Moon and the Sun; further observations being made on the 11th, 15th (distance from 'Cor Leonis', i.e. Regulus) and 19th. His measurement on the 28th, based on the angle between the Moon and the Sun, coincides with the date of Waddington's first reported result but their values are different: Maskelyne recorded a longitude of 29° 44' W, Waddington 29° 33'. Furthermore, they were using different values for the longitude by account, 26° 2' and 26° 44' respectively, so that the differences they noted from the account are themselves different (3° 42' W of account and 2° 49' respectively, the consistency among the figures for the two observers indicating that we are not dealing with a simple clerical mistake).

Throughout the voyage, Waddington's longitudes by account differ from those used by Maskelyne, which is consistent with his remark to Birch that he kept his own account and that the captain had declared it the best on the ship. They were indeed better than Maskelyne's.

A comparison of their longitudes by lunars shows that, after Waddington's first report of 28 February, his positions for a time coincide so closely with those recorded by Maskelyne that they must have been working together or sharing results. This applies to five determinations between 9 and 15 March, including two on the 13th, one using the Sun and the other Regulus. Thereafter they diverge, Waddington's five remaining determinations either differing from Maskelyne's or being taken on different days. A plausible explanation is that, in reporting longitudes by lunars to Birch, Waddington had not discriminated between Maskelyne's and his own observations: this may be the significance of his reference to '*our* Longitudes deduced from Observations' (emphasis supplied), though he could simply have been referring to the ship's company in the plural.

Waddington also told Pigott of his attempts to find longitude by lunars on the voyage, involving Maskelyne in his account and revealing something of his future intentions:

> The Rev. Mr. N Maskelyne & I made several Observations viz 18 In Order to find ye Longt^{d} of ye Ship on our Voyage & finds a Practical & Certain Method & may be depended upon to one Degree of Longitude, we make use only of a Hadleys Quad^{t} & Stop watch for this purpose, but I find ye Quad^{t} wants a little addition wch at my Return I shall do, & also Compute Tables for ye more ready bringing out ye Long^{t} as it requires now 6 Hours of Tedious Computation, & by New Tables may be Computed in 1 Hour.

So it seems that Waddington already had a plan and that he saw himself as an important player in the development of the lunar-distance method for longitude.

Waddington also gave Pigott an account of his and Maskelyne's activities after disembarking. A month of fine weather 'in the Valley' produced some good observations but conditions had been poor in the observatory which had by then been established on a hill about half a mile high. Here Waddington was stationed with the transit instrument and clock, while Maskelyne had the ten-foot sector about two-and-a-half miles distant in the valley, visiting his assistant every two or three days.

Waddington explained that he would leave on the first available ship following the transit but that this could be two months afterwards. In the event, the *Oxford*, an East Indiaman under the command of Captain William Webber, arrived at St Helena on the day of the transit (6 June) and this was the ship that brought Waddington home.[20] Some time was spent repairing a leak but on 29 June they set sail in the company of six other vessels, with the energetic Captain Marriot Arbuthnot (later a colonial governor) as Royal Naval escort commodore in the fifty-gun *Portland*.

The first record in Webber's journal of Waddington's presence on board comes on 23 August, when the 'Long^{de}: by Mr. Waddingtons Obs^{n}:' was noted as 35° 40' west of London, and on 2 September Webber noted: 'Long^{de}. from London at Noon by Observ^{n}: of Dist^{ce}. of ye Sun and Moon'. Waddington would later emphasize his independent determinations of longitude by lunars on his return voyage and the journal provides corroborating evidence for this.

In fact, Webber's journal records an even more eventful voyage than the outward one. On 15 July Arbuthnot gave brief chase to an unknown vessel but desisted when he was in danger of losing contact with the following convoy. Curiously the unknown ship did not take the opportunity to escape but, with the convoy 'brought to' (i.e. stationary), she also stopped, maintaining a station some four miles from the *Oxford*, 'and in this Posture She continued ... when the whole fleet took sight of her, upon so Strange a manner of Acting'. Arbuthnot called a conference of all the commanders and there was general agreement that this was a trap. They suspected that lurking nearby was a squadron under the direction of the French commander Anne-Antoine, Comte d'Aché, and they decided not to make the customary stop at an island (presumably Ascension) for fear of being caught at a disadvantage by d'Aché, whose previous exploits during the Seven Years War were well known.

The journal also reveals that the *Oxford* was carrying French prisoners and that a severe storm on 7 September caused a lot of damage in the fleet, with the *Oxford* coming close to losing contact with the other ships. By the time Waddington reached the Downs on 20 September, he had had an eventful experience of life at sea. Maskelyne had remained on the island to work on an attempted observation of the parallax of Sirius, using the zenith sector, something he had proposed adding to the expedition and not part of Waddington's brief. Circumstances therefore brought Waddington and his longitude ambitions home some seven months before Maskelyne.

LONDON AND LONGITUDE

Waddington wrote to Pigott only on 7 January 1762 that he had returned to London on 21 September after a voyage of three months, excusing his delay in writing on account of catching cold on the boat journey from Tower Stairs to Woolwich, a two-month convalescence and getting settled in London 'with my family'.[21] In compensation, the letter (as usual for Waddington) is crammed with content, astronomical as well as personal. There are solar and stellar observations, further details on the lunar eclipse, and an unsuccessful attempt to derive his altitude on the hill from the going of his clock. 'My Spouse' sends her compliments, so we know for sure that Waddington is married.

In fact, a notice of Waddington's arrival had appeared in the *London Evening Post* for 6 October 1761, with information from him on the transit of Venus observations, the longitude determinations he and Maskelyne had made on the outward voyage and, in particular, Waddington's own determinations on the return voyage.[22] On this occasion, 'Longitude Observations of the Ecliptick Disk of the Sun and Moon, or Moon and Stars, did not differ more than 20 Miles with the Longitude of Portland Point in the British Channel'; this, after eleven weeks and five days at sea. The same account appeared in other news-sheets – the *Whitehall Evening Post* and the *Public Ledger*.[23] This was a serious and public announcement of longitude ambition.

It was reinforced by an advertisement in the *Public Ledger* for 4 November announcing a 'New Mathematical Academy' under the direction of Robert Waddington and George Witchell.[24] Witchell would become the First Mathematical Master (in effect the principal) of the Royal Naval Academy, Portsmouth, in 1767, and later the same year a Fellow of the Royal Society. As regards the New Mathematical Academy, however, the advertisement seems to present Waddington as the principal figure.

Instruction was offered on 'Arithmetic, Merchants Accompts, Computation of Exchaages [*sic*], Geometry, Geography, the Use of the Globes, Gauging, Surveying, Algebra, Navigation, Fortification, Gunnery, and every other Branch of the Mathematics', while 'Gentlemen and Ladies' could be instructed in their own homes in any part of London or Westminster. Waddington was described as 'Mathematical Instrument Maker, Land Surveyor, and Teacher of the Mathematics'. The Academy's address was 'Three-Tun-Court, Mile's Lane, Cannon Street, near the Monument'.[25] There was a particular facility for students of navigation:

> Mr. WADDINGTON has a most convenient Platform upon the Top of the Academy, where he intends to instruct his Pupils in all the Practical Methods of making Observations on the Sun, Moon, and Stars, in order to obtain the Latitude, Variation, and Longitude of a Ship at Sea, by Methods made use of by himself and the Reverend Mr. NEVIL MASKELYNE, F.R.S. in their Voyage to the Island of St. Helena, and on his Return from thence.

Waddington then made a bold announcement:

> Mr. WADDINGTON begs Leave to inform the Public that he frequently made four or five Observations in the Space of half an Hour, an Hour, or two Hours, as Opportunities happened, of the Distance of the Sun and Moon, or of the Distance of the Moon and Zo[d]iacal fix'd Star, whose Latitude was nearly the same as that of the Moon's; then each Observation being computed and compared, they seldom differed more from one another than ten or fifteen Miles in Longitude: By which Means the Longitude at sea or Land, may be generally obtained to less than half a Degree, and always to less than One Degree, including the greatest Error of the best Lunar Tables.

This was a very significant claim: half a degree was the threshold for the maximum reward laid down in the Longitude Act.

Waddington's longitude ambitions place his letter to Pigott of 7 January 1762, with its apologies for the delay in writing, in a different light. His reasons for renewing contact may not have derived entirely from friendship. The address he now used for Pigott was Louvain (Flanders) and Waddington wanted to make use of his contacts with Continental mathematicians and astronomers. He repeated the gist of the claims he made in the news-sheets: his 'considerable number' of observations with a Hadley quadrant had put the ship's position only twenty miles in error after a voyage of three months, when the ship's account was out by nearly four degrees (which at that latitude would be closer to 170 miles). He had already made some enquires regarding how this success placed him in respect of a longitude reward:

> I have applied to some of our Longitude Commissioners, & they Give me for answer that I Compleat only what M. Mayer has done towards it, & altho' my Method of Observing & Computing might be allow'd, as I depend on M. Mayer's Lunar Theory & said Gentleman having made a claim for ye Reward, my Compleating it would no ways entitle me to any part of ye Reward but M. Mayer to it therefore I am at a full stop on this head.

No such communication with Waddington appears in the minutes of the Board of Longitude, so his approach probably was, as he said, to some members of the Board, in conversations or correspondence, outside the meetings and not formally to the Board itself.[26]

Waddington was keen to use Pigott's Continental mathematical connections to make contact with Tobias Mayer, with a view to collaboration:

> Provided it be in your Power & thinks proper to acquaint Mr Mayer of Gottingen of this Obstacle ... I should be Glad to have a few Lines from Mr. Mayer on this head, as his Theory & my New Contrived Quadrant [presumably with the 'little addition'

> he thought of on the voyage to St Helena] &c would fully satisfy ye Longt^{d}. act.

He suggested that Pigott might further encourage Mayer by letting him know that James Bradley seems unlikely to do further work on the lunar theory, 'being fast upon ye decline'; and he appealed to Pigott directly, 'if you or any of your friends can do me this favour to Mr. Mayer I shall be for ever obliged to you'.

Ever one for making full use of paper, in another space he reported the disappointing outcome of his and Maskelyne's observations of the transit, on account of the weather, and that he has no word of the two observers who had set out for Bencoolen. It is surely pertinent to the longitude question that he added: 'Mr. Maskelyne will stay this Year at St Helena.' His main topic was how to determine his longitude difference from Pigott, at Louvain, whether by Jupiter satellites or the meridian passage of the Moon, but in closing he reminded Pigott that he would really like him to write to Mayer about the longitude work so that he could open a correspondence.

Pigott replied as early as 29 January and Waddington wrote again on 17 February, telling him of the house he had taken and the 'large Leaden Platform' for observations, together with a 'little Room for an Observatory'.[27] While dealing with the usual fare of astronomical observation and calculation, and adding observations with the barometer and dip circle, his conclusion reveals his increasing anxiety to make contact with Mayer, reminding Pigott of his former entreaty and adding, 'If you do not care to write him on y^{r}. own acc^{t}. be pleas'd to write him for me.'

Waddington received a letter written on 20 April, but clearly without the introduction he wanted, for when he replied on 20 June, he did not leave the Mayer request to the end but decided on an immediate and direct plea:

> I should be greatly oblig'd to you to write Mr Mayer of Gottingen ye Contents of what I related to you, as to Dr Bradley he is like a man in ye Grave, as being in a State of not doing anything, & has not been at ye observatory since ye Transit.[28]

There was word that Nathaniel Bliss, Savilian Professor of Geometry at Oxford, would be his successor. The other news was that Maskelyne had returned, bringing further confirmation of the lunar method for finding longitude at sea. It is clear that at this stage Waddington saw himself acting together with Maskelyne in approaching the Board:

> we are going to lay this Method before ye Lords of the Admiralty & Commissioners for ye Longitude in Order to request a Tryal to be made of it on Board his Majesties Ships &c. In my next I will advise you whether they approve of our offer or no & in ye Interim hope that you will write Mr Mayer.

In another letter that Waddington began late in June but completed some two weeks later, he stated as one of the achievements of the St Helena expedition, 'descovering [*sic*] a General Method of determining ye Longitude at Sea'[29], but we learn from the same letter that he had heard of the death of Tobias Mayer. He also reported a conversation with 'Mr Harrison' (probably William rather than his father John):

> Mr Harrison Yesterday [9 July 1762] told me that he came within y^{e} ½ degree of ye Longitude upon his return with his watch from ye West Indies & that he will make his demand for ye 20 Thous^{d}. L. at ye next Board of ye Comis^{rs} wch will be in a month's time as to what was reported abt his Watch loosing 8½' is true, but he told them before he departed that it would loose a certain N^{o}. of Seconds pr Week, & those being accounted for & allow'd ye Watch did not differ 2', from what he said, it would &c.

The question of whether a rate had been properly stated before the voyage did indeed become a central issue at the meeting of the Board.

It is clear that Waddington was maintaining his well-informed interest in the longitude question and doing what he could in 1762 to be part of the London mathematical scene. A paper of his was read at the Royal Society on 14 January, containing barometer and thermometer readings at different locations on St Helena, with the thermometer series continued throughout the voyage home.[30] In a conclusion he added that, on St Helena, Maskelyne and he had observed 'the bright Star in the foot of the Centaur' (Alpha Centauri) and found it to be a binary with 'a Sensible Parallax', whose component stars vary in their mutual distances. These were observations of potential importance for cosmology but, before the work of William Herschel, this was not a topic of widespread interest. Waddington subsequently attended a number of Royal Society meetings in 1762, introduced by one of the Secretaries, since he was not a fellow.[31]

There seems to be no evidence that the Board of Longitude was inclined to associate Waddington with any reward for progress with lunars and, for the time being, his academy offered his most realistic possibility of making a living from mathematics. A further bold move was his publication, over two issues of Benjamin Martin's *Miscellaneous Correspondence* for June and July 1762, of an entire determination of longitude from the distance of the Sun and Moon, taken with a Hadley quadrant on board the *Oxford* on 2 and 4 September 1761 by 'R. Waddington, Teacher of Mathematics at the Mathematical Academy in Three Tun-court, Mile's-lane, London'.[32] He made use of the *Connaissance des Temps* for tables of the Sun and Moon and, claiming that his measurement was closer to the truth than the ship's account, he concluded that 'By this method, the Longitude of a Ship at Sea may generally be determined to less than half a Degree.' Further, in a third part of his paper, published in the August issue, he gave a determination made by a commander of an East Indiaman at his Mathematical Academy, this one taken on land (perhaps from the academy's observing platform) and with an error of only three miles.[33]

A letter to Pigott of 25 October provides further evidence that Waddington was teaching lunars at his Academy.[34] He asked Pigott to write to Lalande for half a dozen copies of the 'French Ephemerides' for 1764 (the *Connaissance des Temps*, then edited by Lalande), to be used by captains of East India Company vessels. Indeed he affirmed that, if he could have a regular supply of the ephemerides two or three years in advance, he could sell 'a considerable N^{o}.' to these officers who 'would take them in Order to make use of for finding the Long^{t}. at sea by ye dif. of Long^{t}. of ye sun & moon'.

This led him on to updating Pigott on the Harrison side of the longitude story. He related that William Harrison had returned from Jamaica, since when a further £2,000 had been paid to John Harrison, making a total reward of £5,000. William was to go to the West Indies again in the coming April, for a final trial, when £1,000 would be paid should the watch fail, or a reward paid for success according to the terms of the Longitude Act. 'Hence', he wrote, 'I find little hopes of any part of the reward to fall to ye Lunarian Method.'

It may have been a realization that the longitude case was going against lunars, as well as his relative invisibility to the Board, that nudged Waddington further towards pursuing instead an independent and private route to making commercial capital from the lunar method. With all hope of collaboration with Mayer gone, Waddington's concluding news announced a different initiative: 'I am very busy Calculating & Compiling a Book of Practical Navigation wch. I hope to have Publish'd by ye 10th of Jany I should be glad to send you provided I know how.' As always, Waddington found astronomical observations and calculations to share, at times still adopting the tone of a sympathetic teacher. On this occasion he included problems with the observations of Lalande, in this matter also associating himself with Maskelyne: 'Mr. Maskelyne & I am of opinion that the French Mathematicians has had ye Misfortune to fall under a like defect with their Sectors by ye Plumb lines as he did with his at St Helena.'

Waddington's book was published in March 1763 but before that he would make a desperately unsuccessful bid for a close engagement with the Royal Society – not for Fellowship but employment. On 11 January 1763 the younger Francis Hauksbee died. He had been clerk and housekeeper to the Society for some forty years and, as they moved to appoint a successor, the Society had first to settle and formulate all the roles that his duties had come to encompass, such as looking after their Repository. When Council met on 20 January with this as their 'immediate business', there were already five applications. Read to the meeting in the order of receipt, the first was from James Ferguson, lecturer and author, and the third was from Robert Waddington.[35]

When Council met again a week later, there were six candidates, all in attendance, and the record of the precedent set by Hauksbee's appointment of 1723 had been consulted. Each candidate was to be examined with respect to six criteria: handwriting, languages, natural history, mathematics and mechanics, books in libraries, and provision of financial security. They were called in separately and their answers noted, and the record adds to our knowledge of Waddington.[36] His handwriting was fine but in languages he could say only that he had done some teaching of Latin. Mathematics and mechanics were covered, of course, since 'He teaches divers Parts of practical Mathematics; and hath for some time followed the Business of an Instrument Maker' and he had been assistant to Maskelyne on St Helena. Outside his professional sphere, however, he was less secure: 'He says he has read & understands the Philosophic Chemistry of Boerhaave, and hath analysed divers particular natural Bodies; but seems not greatly acquainted with what is called natural History.' As regards the management of books, Waddington said that, when he was clerk to a merchant at Hull, he had had charge of his library. The candidates each had to state their family circumstances and Waddington had a wife and three children, the eldest nineteen and the youngest nine. He 'believes' he could find security of £1,000.

On 3 February a full meeting of the Society settled down to what would be a long session to determine the appointment.[37] The petitions of all six candidates were read, followed by a report on their examination by the Council. Each candidate was called in separately and asked whether they wanted to add anything to their examination. Waddington, realizing his weakness in natural history, said that 'he had formerly been acquainted with Botany'. The ballot required every Fellow voting to come forward in a separate operation for each candidate and, having been given a ball, to cast it as a positive or negative vote. Eighty-four such votes were cast in respect of the first candidate, Ferguson, and eighty in respect of the others: presumably four fellows had quickly grasped the implications of this procedure and decided they were urgently needed elsewhere. After four candidates had been balloted, one, Emanuel Mendes da Costa, was so far ahead (62 for, 18 against; Ferguson was lying second with 33 for and 51 against) that Maskelyne seems to have contrived to persuade the last two that their causes were hopeless and they 'declined giving the Society any further trouble'. Poor Waddington had come an abysmal last, with seven votes for and seventy-three against. If natural history was the key criterion, Mendes da Costa was the man for the job, but the appointment was a disaster as he was dismissed in December 1767 for embezzling Society income, and then imprisoned for fraud.

Perhaps Waddington realized at last that he was not likely to succeed in the mainstream mathematical establishment and would have to live by his commercial wits, such as they were. He would make one more attempt at an official appointment but, as we shall see, that was no more successful than with the clerkship. Still, there was the book of practical navigation to see through the press. It was announced in *Lloyd's Evening Post* as published 'This Day', on 4 March 1763, with the title *A Practical Method for Finding the Longitude and Latitude of a Ship at Sea, by Observations of the Moon*. The author was Robert Waddington, 'Teacher of the Mathematicks'.[38] A quarto volume of some seventy pages, it represents a great deal of work and of investment in the lunar method by Waddington (Fig. 2).[39]

If this was, in a sense, a 'private' investment, Waddington had not lost all hope of patronage from the Board of Longitude: he addressed his treatise to the First Lord of the Admiralty 'And the Rest of the Commissioners for the Longitude'. In his introduction he naturally moved swiftly to establish the successful practical experience of the voyage to St Helena 'by the Rev. Mr. NEVIL MASKELYNE, F.R.S. and myself', and of the return voyage in the *Oxford*, 'by myself'.[40] He claimed to have made considerable improvements to the method: as regards observing, by taking the average of several observed distances and, as regards computing,

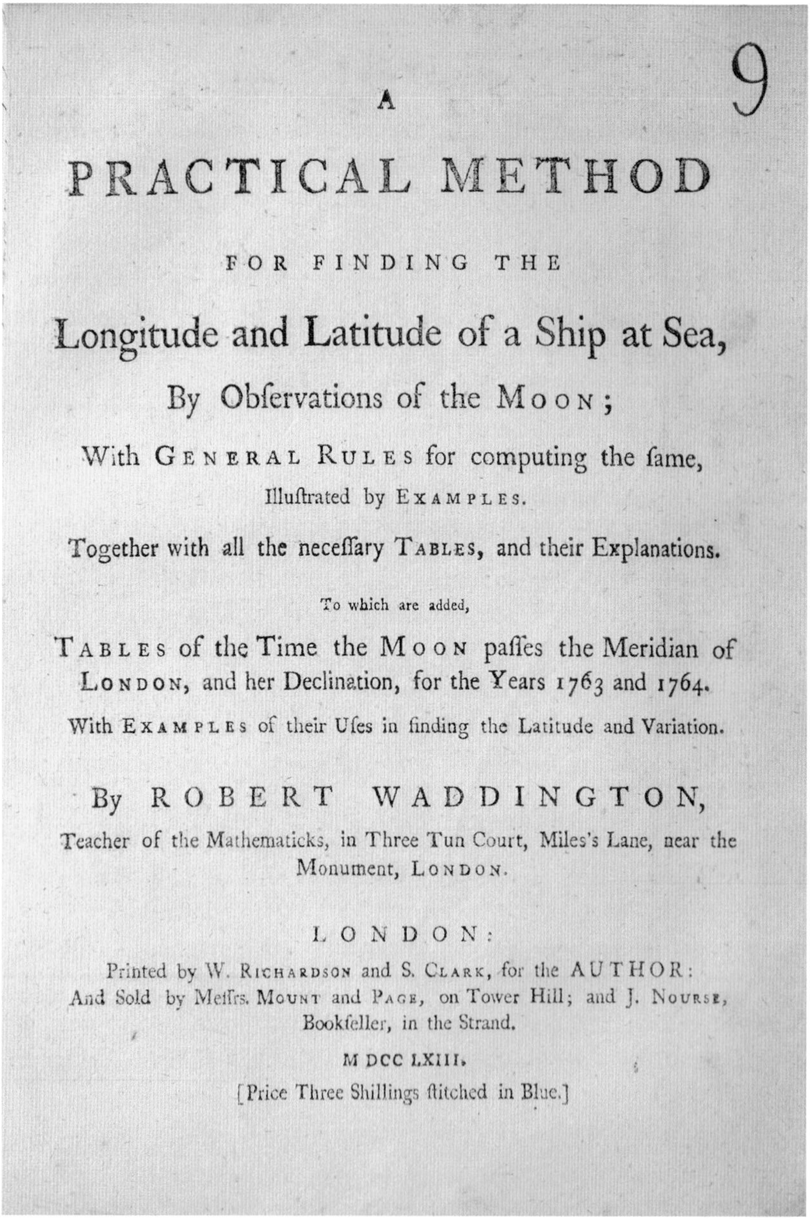

9

A

PRACTICAL METHOD

FOR FINDING THE

Longitude and Latitude of a Ship at Sea,

By Obſervations of the MOON;

With GENERAL RULES for computing the ſame,

Illuſtrated by EXAMPLES.

Together with all the neceſſary TABLES, and their Explanations.

To which are added,

TABLES of the Time the MOON paſſes the Meridian of LONDON, and her Declination, for the Years 1763 and 1764.

With EXAMPLES of their Uſes in finding the Latitude and Variation.

By ROBERT WADDINGTON,

Teacher of the Mathematicks, in Three Tun Court, Miles's Lane, near the Monument, LONDON.

LONDON:

Printed by W. RICHARDSON and S. CLARK, for the AUTHOR:
And Sold by Meſſrs. MOUNT and PAGE, on Tower Hill; and J. NOURSE, Bookſeller, in the Strand.

M DCC LXIII.

[Price Three Shillings ſtitched in Blue.]

Fig. 2: Robert Waddington's 1763 navigation manual, *A Practical Method for finding the Longitude and Latitude of a Ship at Sea, by Observations of the Moon.*

by providing a table giving 'the variable Quantity of the Moon's Motion by Interpolation, whereby I could readily get the true Longitude and Latitude of the Moon from the Paris Ephemeris, to any Day, Hour, and Minute required'. The whole calculation could be completed in about three-quarters of an hour, enabling him to perform separate computations on three or four observations and to find the mean. He claimed his results 'may always be depended upon to about a Degree, and generally to less than Half a Degree, or One Quarter thereof'. After such a claim he need hardly have added: 'I cannot but earnestly wish it may meet with the Encouragement due thereto.' In this context 'encouragement' was understood to include financial reward.

It is confirmed in the introduction that Waddington had indeed been instructing ships' officers of the East India Company. He added that he had computed tables for their use for January and February 1763, and he included these 'by way of Specimen', along with the general rules for taking the observations and making the calculations. It was in the area of future lunar tables that he saw room for improvement:

> if Tables in the like Manner were computed and published for Three Years before-hand (i.e. for Three Years to come) and from the best Astronomical Tables, with Mr. MAYER's corrected Equations applied, and the commanding Officers of Ships properly instructed, that after a few Years (I dare venture to say) no other or better Method of determining the Longitude at Sea would be desired.

We learn from his treatise that Waddington sold Hadley quadrants at prices from £4 5s 6d to 7 guineas, depending on whether they have a telescope and a vernier. An additional 15 shillings would secure Waddington's own design of fitting or mount to facilitate holding them in the variety of oblique positions that lunar observations required.

Waddington included an example of an observation and calculation made on board the *Oxford*, based on measurements of the distance between the Sun and Moon; taken on 2 September, this was one of the dates noted in the captain's journal. He was in latitude 40° 8' N and derives a longitude of 2° 44' west of London. In a second example from the *Oxford*, on 8 July in latitude 7° 50' S, 'assistants' – presumably members of the crew – had taken the altitudes of the Sun and Moon (not reliably, as he demonstrated). The calculated longitude was 21° 42' west of London.

Waddington followed this with observations of distances between the Moon and stars taken on the *Prince Henry* (on the outward voyage with Maskelyne) and a further example taken on St Helena on 14 June, using the star Spica and deriving a value of 5° 43' west of London, which he said was in error by eleven miles (the modern value is 5° 42' 0" W). The remainder of the book is filled with lunar and solar tables, instructions for their use, a table of refractions, and various elements of astronomy relevant to navigation.

The *Monthly Review* for October 1763 reviewed *A Practical Method* over four pages with the running head 'Waddington's Method for finding the Longitude and Latitude at Sea'.[41] The well-informed reviewer placed the subject in the context of the Board and the longitude reward. Waddington was judged to have improved the prospect of the lunar method, 'and from the account he has given us in this work, there is great reason to hope, that his success will animate others to put the method in practice, as it cannot fail of proving of the utmost importance to Navigation.' Add to this, the writer continued, that the observations were made with the Hadley quadrant, that Waddington's directions for the necessary calculations were plain and included tables for shortening the operation:

> We may therefore congratulate our countrymen, that by the method here explained, and the accurate time-piece constructed by the ingenious Mr Harrison, (a full Account of which will shortly appear in our Review) the great Desideratum in Navigation, the discovery of the Longitude, will be soon compleated, and consequently the art itself reach its ultimate perfection.

The story is usually told with Harrison and Maskelyne as protagonists for the rival methods for finding longitude; here we are offered Harrison and Waddington.

It was also in 1763 that Waddington made direct contact with Lalande, who was visiting London. Waddington wrote to Pigott on 18 April that he had spoken to the distinguished French astronomer twice at the Royal Society and that he would come to dinner that very week.[42] Lalande's diary records a Royal Society meeting at Crane Court on 17 March, only his third day in London, following dinner with the members of the Society's Club, including Maskelyne. At the Society he met Waddington, 'who was in St Helena with Maskelyne and who has a school near the Monument'.[43] That Lalande did in fact visit Waddington's academy is confirmed by a subsequent letter (18 January 1764) to Pigott:

> The French Academicians, I see several times at ye meeting of ye Royal Society, particularly M. de la Lande, s^{d}. Gent. gave me a call at my House & I made him a pres^{t}. of one of my Treatises on ye Long^{t}. I shew'd him some mistakes in ye Ephimerides, wch. he promis'd sho'd be corrected.[44]

Waddington moved on to the forthcoming solar eclipse ('I have already engag'd to go to Canterbury or Dover where I make it pretty much Annular') but returned to the French Academicians in a postscript. In addition to Lalande, they were the mathematician, Charles Etienne Camus, and the horologist Ferdinand Berthoud, who had been sent by the Académie des Sciences in the expectation of examining Harrison's timekeepers. As is well known, Harrison would not show them the watch undergoing rating trials organized by the Board of Longitude and, as usual, Waddington is well informed:

> Harrison's Watch goes upon ye Final Tryal in 3 Weeks time, it is now under examination of its going, and I am inform'd it looses 1,5" p day regularly. ye owner would not shew it to ye French Academicians those Gent. got made F.R.S. before they departed.

Waddington used his solar-eclipse predictions to maintain visibility for himself and his academy by publishing a paper in Martin's *Miscellaneous Correspondence* for September 1763. This included his conclusion that the eclipse, to occur on 1 April 1764, would be annular along the south coast of England.[45]

A Supplement to the Treatise for Finding the Longitude of 1764 was Waddington's last publication from the address in Three Tun Court.[46] It included updated tables for 1764 and a promise of a further supplement for 1765. When he wrote again to Pigott, on 27 March 1764, William Harrison had left for Barbados and Waddington had moved to Rolls Building in Chancery Lane.[47] He did not explain the reason for the move but it may have been connected to the lease, as he later wrote that it took place on Lady Day, the customary end-date for such contracts.[48]

We might take these events to signal the effective end to Waddington's pretensions to a longitude reward. For a time he had some success in forging a public association with the lunar-distance method, alongside Maskelyne. In an extensive, and negative, account of Harrison's chronometer and its trial in Robert Heath's *The Palladium-Supplement, Enlarged* (for 1764), the author says:

> The Rev. Mr. Maskelyne (one of the Observers lately sent to the West Indies), and Mr. Waddington, of London, (sent to observe the late Transit of Venus,) have both treated on these Subjects, of observing the Longitude and also Latitude at Sea by the Moon.[49]

However, this account did not reflect Maskelyne's view of the matter, which had no conspicuous place for Robert Waddington. In the month following the publication of *A Practical Method*, Maskelyne published his own *British Mariner's Guide*, also dedicated to the Longitude Commissioners (Fig. 3).[50] He, too, based his claims for a practical lunar method on his voyages to and from St Helena but Waddington was not mentioned as a partner in the project.

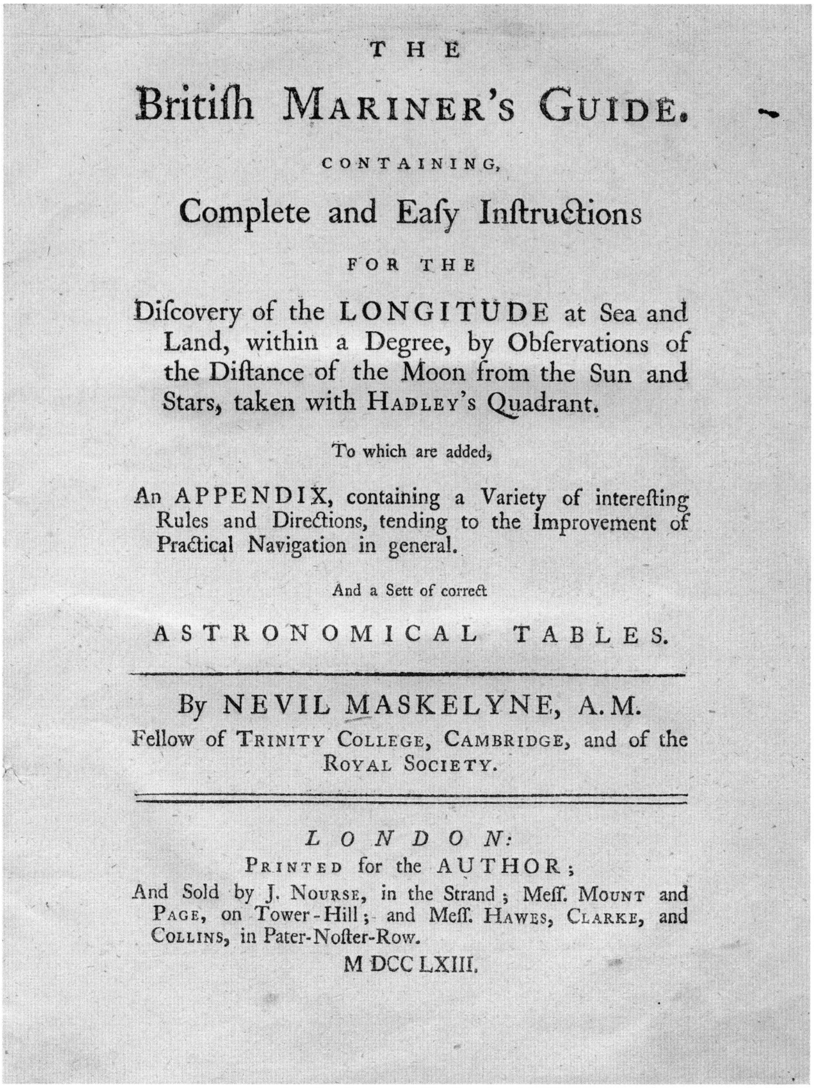

THE

Britiſh MARINER'S GUIDE.

CONTAINING,

Complete and Eaſy Inſtructions

FOR THE

Diſcovery of the LONGITUDE at Sea and Land, within a Degree, by Obſervations of the Diſtance of the Moon from the Sun and Stars, taken with HADLEY'S Quadrant.

To which are added,

An APPENDIX, containing a Variety of intereſting Rules and Directions, tending to the Improvement of Practical Navigation in general.

And a Sett of correct

ASTRONOMICAL TABLES.

By NEVIL MASKELYNE, A.M.
Fellow of TRINITY COLLEGE, CAMBRIDGE, and of the ROYAL SOCIETY.

LONDON:
PRINTED for the AUTHOR;
And Sold by J. NOURSE, in the Strand; Meſſ. MOUNT and PAGE, on Tower-Hill; and Meſſ. HAWES, CLARKE, and COLLINS, in Pater-Noſter-Row.
M DCC LXIII.

Fig. 3: Nevil Maskelyne's *British Mariner's Guide. Containing Complete and Easy Instructions for the Discovery of the Longitude at Sea and Land … by Observations of the Distance of the Moon from the Sun and Stars*, published in 1763.

WADDINGTON'S NAVIGATION

Things seemed to be looking up for Waddington in August 1764. His new address was the Royal Academy, Portsmouth, the Admiralty's school for training aspirant naval officers, where he moved on 9 July, and he could tell Pigott that he had been appointed Second Master.[51] The salary was a very respectable £100 and came with a house, a brew-house, cellars, a garden, a garden house, a stable and a chaise house – somewhere to keep his carriage, should he ever have one. His family were all present and, not surprisingly, he said 'I hope pretty well settled'. Harrison was back from Barbados, and the news was that the watch varied by only 54 seconds over both voyages, 'therefore he must be in a fair way for getting y^{e} reward for y^{e} $Long^{t}$.'.

Waddington remained very well informed about the progress of Harrison's claim for the longitude reward, giving Pigott a full and accurate account of his dealings with the Commissioners in a letter of 20 March 1765.[52] He had a clear grasp of the Board's insistence that Harrison must 'discover' the watch, that is reveal its construction and the principles of its timekeeping. The full reward could be made only 'when he has made a discovery of ye principles upon which his Timekeeper is constructed by means whereby others may be made'. Some of the detail suggests that Waddington was still in contact with Maskelyne or someone else close to events: 'I am inform'd that his Watch is not perfect, & that he has not overcome ye effects of heat & Cold, as by him reported – some days it wo'd vary 1 or 2 & other 5 or 6".'

In fact, Waddington wrote in a letter of 3 June 1765 that he had recently dined at the Royal Observatory and reports also that he was to attend the Admiralty and 'to Compute 8 different Longitude Observations made at sea'.[53] However, something was beginning to go wrong in Portsmouth – 'something', he tells Pigott obscurely, 'material in respect of my station & situation in ye Academy'. He was still in post in December and was able to tell Pigott that Harrison had been awarded £7,500, making a total so far of £10,000.[54] By the end of 1766, however, he had left – 'things there not being in an agreeable manner, but very much otherwise'.

While Waddington wrote, 'I took leave of ye Royal Academy at Portsmouth', other sources indicate that both masters had been dismissed.[55] Little is known about the circumstances and the First Master, John Robertson, went on to have a successful tenure as Clerk and Librarian to the Royal Society, replacing the disgraced Mendes da Costa in 1768 (Waddington did not apply again).[56] The new First Master was George Witchell, Waddington's original associate at his Mathematical Academy, while Waddington himself moved his operations to what would later become a famous address: Downing Street, Westminster: 'I have taken a lease of a House & Garden situated near ye Park, my principal business is Surveying, Planning, & valuing of Land, in wch. employ I propose to establish my two sons – I teach privately & read philosophical Lectures.'

The letter bearing this news was begun in December 1767 but there must have been delay in sending it, for in an addition dated 5 February 1768, Waddington dealt with the forthcoming transit of Venus, the second of the eighteenth-century pair of transits. He said that the Royal Society were planning two expeditions – one to South America, to the Straits of Magellan, the other to Hudson's Bay – and that he had been 'nominated' for one, 'but I cannot as yet say whether I dare Venture to spend a Winter in ye Could Climate of Hudson's bay'.

This account is scarcely consistent with the record at the Royal Society, where a committee had been established in July 1767 to plan for the 1769 transit.[57] It included Maskelyne, Ferguson, the telescope maker James Short, and the astronomer John Bevis, who, on the basis of the letters to Pigott, Waddington clearly considered a friend. Short, Maskelyne and Bevis each produced a list of some half-dozen suitable candidates, but no one nominated Waddington. He did, in fact, apply, as letters offering to go to Hudson's Bay from him and from Joseph Dymond, former assistant observer at Greenwich, were read at a Council meeting on 21 January 1768.[58] Council proceeded directly to offer the Hudson's Bay station to Dymond and William Wales, who was assisting Maskelyne with computations for the *Nautical Almanac*. Nothing suggests that Waddington was seriously considered and the matter was settled before he wrote to Pigott of his 'nomination'. We might imagine a problem connected with the Portsmouth episode or note simply that the men appointed were respectively some twenty or ten years younger than Waddington, but it is hard not to conclude that he was no longer taken seriously in the influential circles of astronomy and navigation.

In 1771 Waddington did, nonetheless, make his one formal approach to the Board of Longitude – not in connection with a longitude method but related to his now more diversified engagement with navigation. On 11 May, with Maskelyne present, the Board received and read a petition from Waddington to the effect that he had a greatly improved model for 'the common Binacle or Steerage Compasses now in use'.[59] He claimed that his instrument had been successfully tried on many coastal vessels and he enclosed a certificate of approval signed by officers, including the commander, of the Royal Naval sloop *Zephyr.* He asked for trials on warships, 'that he may be rewarded, as the merits of his improvement shall appear to deserve from the Certificates of the Commanding Officer of such a Ship'. Waddington was advised that he would have to apply to the Admiralty.

Waddington was also active in improving the azimuth compass for measuring magnetic variation at sea. When on board the *Prince Henry*, ten years earlier, he would have become aware of the care and application Captain Haggis gave to his frequent determinations of magnetic variation from measurements of solar azimuth or amplitude (or both), by himself or with the first or second mate (or both) working with different compasses at separate stations on the ship. These were for up-to-date corrections to the steering compass and for adjustments to dead reckoning, but they were also used for position-finding in the absence of a reliable longitude fix.

Several further mathematical works were published by Waddington. An outline account of common surveying instruments, with some elementary geometry and instructions for land survey, appeared in 1773.[60] Its purpose seems to have been that of a useful handbook for pupils. A more ambitious treatise on the sextant followed in about 1775.[61] After a practical account of the instrument, Waddington explained again how to find longitude by lunars, now using such recent resources as the *Nautical Almanac*, and he appended general accounts of refraction, parallax and the solar system. His value for solar parallax derived from observations of the 1761 transit of Venus and his examples of longitude determinations still included measurements from the *Prince Henry* and the *Oxford*.

An Epitome of Theoretical and Practical Navigation, appearing in 1777, was a serious attempt at a comprehensive navigational textbook, 'containing' the title page affirms, 'a complete system of that art'.[62] This was a large work (241 pages) and cannot be treated fully here, yet Waddington was not content and in the following year published a forty-five-page supplement, *The Sea Officers Companion, Being an Appendix to Waddington's Navigation*.[63] The author of the *Companion* was there described as 'Teacher of Mathematics, Surrey-Side, Westminster-Bridge.'

In one of Waddington's later advertisements he offered accounting and book-keeping services, legal and contractual advice, and surveys and valuations, alongside the more usual instruction in a range of mathematical arts.[64] He had reported surveying 'in ye Country' to Pigott in May 1768.[65] Further letters relate to the sale of James Short's effects, and to Jeremiah Sisson making a mural quadrant for an observatory to be built for the King in Richmond Gardens, as well as a one-foot quadrant for Waddington himself, for use in surveying.[66] He observed the 1769 transit of Venus successfully at Paddington: 'both ye Contacts were universally seen ab^t. London'.[67]

Waddington's versatility and his relentless search for profitable mathematical work led him also to publish on the practice of gunnery at sea. The only example of his practical, self-styled 'elementary' tract that I have found dates from 1781, which would make it posthumous, but in the preface to his own more learned treatise on gunnery of 1779, the fractious Reuben Burrow described Waddington's publication as 'contemptible'.[68] Waddington had at least had seaborne experience of seeing gunners at work, since on the *Prince Henry* Captain Haggis had 'exercised' the guns occasionally to keep them, and his gunners, in trim.

As he wrote the last letter to survive in the series addressed to Pigott, dated 29 May 1775, Waddington was in Dunkirk anticipating a trial of his new azimuth compass, particularly adapted to bad weather, and enthusing over his new 'Astronomical Theodolite': 'Mr Ramsden [the instrument maker] says that very few Astronomical Quadrants will perform so well.'[69] So, in spite of all his setbacks, there is little sign of Waddington giving up on his life of mathematical enterprise. Perhaps in anticipation of the hazards of travel abroad, he had made a will on 12 May 1775. Robert Waddington 'of Downing Street Westminster' bequeathed his estate to his wife Margaret as sole executor. The will was proved by the oath of his widow on 29 October 1779.[70]

The story of the search for a longitude method in the eighteenth century has generally been written as a contest between its most celebrated protagonists, Harrison and Maskelyne, with other efforts scarcely taken seriously. The life of Robert Waddington and his struggle to make the lunar method work for him – if not by the official route then by a private, commercial one – shows how the longitude and its Commissioners could weave their influence through the varied work of a more minor mathematical player. Waddington's high-water mark came early in his career and defined much of what he sought to achieve from then on. He did everything he could to make the voyages to and from St Helena the springboard for much more, but he failed.

Waddington's work on lunars mirrored that of Maskelyne: his *Practical Method* was the equivalent of Maskelyne's *British Mariner's Guide*; his *Supplement* the counterpart to the *Nautical Almanac*. Maskelyne was able to achieve an official standing for his programme, whereas Waddington's remained a private, commercial enterprise. Maskelyne commanded all the resources that Waddington lacked – education, official status and, we might imagine, a greater intellectual and organizational talent for prosecuting the lunar method. There was no role in Maskelyne's plan for his sometime assistant but neither is there evidence that he deliberately sabotaged Waddington's ambition for some form of joint action or partnership. Such an idea may have been so irrelevant to Maskelyne that he was scarcely aware of it: the most explicit statements we have on their relationship are those we have seen in Waddington's private letters to Pigott.

Eventually Waddington did decouple the reputation he sought to promote from that of his superior. The identity he assumed for his *Epitome of Theoretical and Practical Navigation* (1777), on the title-page of his supplementary *Sea Officers Companion* in the following year, was 'Waddington's Navigation': that is, a standard textbook firmly identified with its author. In both the *Reflecting Sextant* and the *Epitome*, Waddington was still drawing on the observations made at sea some sixteen years previously but he no longer mentioned Maskelyne. The *Epitome*, according to its title-page, was: 'founded on many Years Experience in the Several Voyages made by the Author; more particularly, that to St. Helena, when appointed by the Royal Society, to observe the Transit of Venus over the Sun.'

FURTHER READING

The other half of the Waddington-Maskelyne relationship is best approached through Derek Howse's biography, *Nevil Maskelyne: The Seaman's Astronomer* (Cambridge: Cambridge University Press, 1989). Biographical accounts of Waddington himself, however, are almost non-existent. He has an entry in E.G.R. Taylor, *The Mathematical Practitioners of Hanoverian England 1714–1840* (Cambridge: Cambridge University Press, 1966), pp. 219–20, and will appear in the revised edition of Thomas Hockey, et al., eds, *Biographical Encyclopedia of Astronomers* (unpublished at the time of writing), in an entry by Paul Murdin. He is listed in R.V. and P.J. Wallis, *Index of British Mathematicians. Part III. 1701–1800* (Newcastle-upon-Tyne: Project for Historical Biobibliography, 1993) and in Peter

Eden, *Dictionary of Land Surveyors and Local Map-Makers of Great Britain and Ireland, 1530–1850*, 2nd edn by Sarah Bendall (London: British Library, 1997).

Within the larger story of the eighteenth-century search for a method for finding longitude at sea, the most ambitious work on the subject to date is William J.H. Andrewes, ed., *The Quest for Longitude* (Cambridge, Mass: Harvard University, 1996). It is a stark reflection of the restricted focus of recent interest that, notwithstanding the book's detailed treatment of many things, Waddington is not so much as mentioned, in spite of the substantial record he left and which it is almost impossible not to stumble across without even looking. Attitudes are surely bound to change with the recent availability of the rich online resource from Cambridge University Library of the papers of the Board of Longitude, http://cudl.lib.cam.ac.uk/collections/longitude.

The middle-ranking, jobbing mathematicians, who made a living through teaching, writing, publishing, surveying, instrument-making and perhaps occasional paid astronomical work, are well covered in the *Oxford Dictionary of National Biography*, where the likes of Mason, Dixon, Bevis, Bird, Ferguson, Witchell, Martin, Robertson, Wales, Short, Sisson and Burrow, all mentioned above, can be found – but not, of course, Waddington.

CASE STUDY B

THE PROJECTS OF EIGHTEENTH-CENTURY ASTRONOMY

Maskelyne's voyage to St Helena demonstrated his interest in and commitment to many of the most significant projects of practical astronomy in the eighteenth century. These projects were engaged with problems that developed out of Newtonian natural philosophy and from the interests of maritime nations. Some were more obviously practical issues than others, but even the most apparently abstruse-sounding of them were intimately linked to perfecting astronomical tables and the use of astronomy for navigation and cartography. Thus, both the grand nature of problems that were about understanding the shape, scale and motions of the heavens, and the potentially useful spin-offs from solving them, could be – and were – used by astronomers across Europe to gain patronage from institutions, monarchs and governments.

In Maskelyne's case, the voyage was one of two organized as an initiative of the Royal Society, with the backing of King George II and assistance from the Admiralty and the East India Company. Both missions were the rather tardy British response to the forthcoming transit of Venus, a rare event, when Venus can be seen to cross the face of the Sun.[1] Other nations, particularly France, were already organizing expeditions to make observations of the transit, and thus national pride was also at stake in making sure that the British were involved in what was to become a huge, multi-national undertaking.

The importance of transits of Venus lay not just in their rarity but because they were a means to measure what Maskelyne called in his 'Autobiographical Notes', 'that great desideratum in Astronomy', the Sun's parallax.[2] A measurement of parallax, or the slight apparent shifting of the position of a distant object in relation to a less distant one as viewpoint is changed, is a way to calculate how far away the former is. This meant that, if the transit were observed and timed at widely separated positions on Earth, it would (at least in theory) be possible to work out the absolute distance between the Earth and the Sun. This was a fundamental measurement, later known as the Astronomical Unit, from which the distances to other planets, previously known in relative rather than absolute terms, could also be derived. Having a better knowledge of the distances and masses of bodies in the Solar System would improve the theory of their motions, particularly the tricky matter of the Moon's, affected as it is by the gravitational pull of both Earth and Sun.

The observations needed to be taken from as widely separated geographical latitudes as possible, taking into account the path of visibility of the transit. St Helena is today given as 15° 56' S and Bencoolen, now Bengkulu (Sumatra), the intended location of the other British expedition, is at 3° 47' 44" S. These observations would be compared with the partial transit observable at Greenwich (51° 28' 38" N) and any others communicated by individuals or official institutions and expeditions across the world. The crucial measurements were to establish the precise different local times at which the planet's ingress (entrance) and egress (exit) over the face of the Sun were observed.

When the Royal Society requested support from King George II in 1760, they emphasized patriotism. They pointed out that it was an Englishman, Edmond Halley, who had suggested back in 1691 and 1716 that the future transits of 1761 and 1769 could be used to determine the distance of the Sun. As the Society told their King, 'it might afford too just ground to foreigners for reproaching this nation in general (not inferior to any others in every branch of learning and more especially in astronomy)' if the French and Russian court were to send observers on their planned global expeditions, but the British did not.[3]

Official support was soon forthcoming 'for the promotion of a Science so intimately connected with the Art of Navigation as well as for the Honour of the Nation'.[4] As described in the Introduction, Maskelyne was quick to insert himself into the process of planning and organization. At James Bradley's request, he provided the Royal Society with a list of instruments that would be required. Each expedition should be supplied with a two-foot reflecting telescope, a micrometer by Dollond and a clock with a compound pendulum (Fig. 1).

Two additional instruments were to be ordered for the St Helena expedition: an eighteen-inch astronomical quadrant and a ten-foot-radius zenith sector. These would aid Maskelyne in the other projects that he proposed to undertake, after the transit observations, during a year-long stay on the island.

Fig. 1: A portable reflecting telescope by Francis Watkins, c.1780, of a similar type to those taken on voyages of exploration, although this example was probably used by an amateur. The preferred maker was John Dollond, but James Cook took a telescope by Watkins on his 1769 voyage to observe the second of the pair of eighteenth-century transits of Venus.

The most significant of these additions had been described by Maskelyne in a paper presented to the Royal Society on 26 June 1760. It contained 'A proposal for discovering the annual parallax of Sirius'.[5] This would involve observing that bright star, which passed almost overhead at St Helena, through the course of a year. Observing at the zenith removes the problem of atmospheric refraction, allowing very precise measurements to be made. It is for that kind of work that a zenith sector telescope is designed. Extreme precision was required for this observation because stellar parallax, from which the distance from the Earth to a star might be derived, would be a very small measurement indeed, reflecting the great distances involved. If successful, this observation would not only give some sense of the scale of the Universe but it would also provide proof that the Earth orbits the Sun.

While the so-called Copernican theory of the Sun-centred system was widely accepted by this date, incontrovertible observational proof was still a missing piece of the cosmological jigsaw. Bradley had observed the Earth's annual motion indirectly when he established the phenomenon of the aberration of starlight with his own zenith sector (Fig. 2) but, as Maskelyne wrote in his paper, it was only the annual parallax of a fixed star, caused by the Earth's orbit, that 'would be the fullest and directest proof of the Copernican system'.[6]

Many had previously tried to observe stellar parallax, including Bradley himself, but Maskelyne argued that using the best instruments available, together with the St Helena location, gave the best chance yet for success. His idea had been inspired by his examination 'of the observations of the zenith distances of Sirius, taken at the Cape of Good Hope, in the years 1751 and 1752 by that excellent astronomer, and diligent observer, the Abbé de la Caille'.[7]

As it happened, Maskelyne failed in the two official tasks with which he had been entrusted and he stayed ten rather than twelve months on St Helena. The observation of the transit of Venus was clouded-out and the measurement of the parallax of

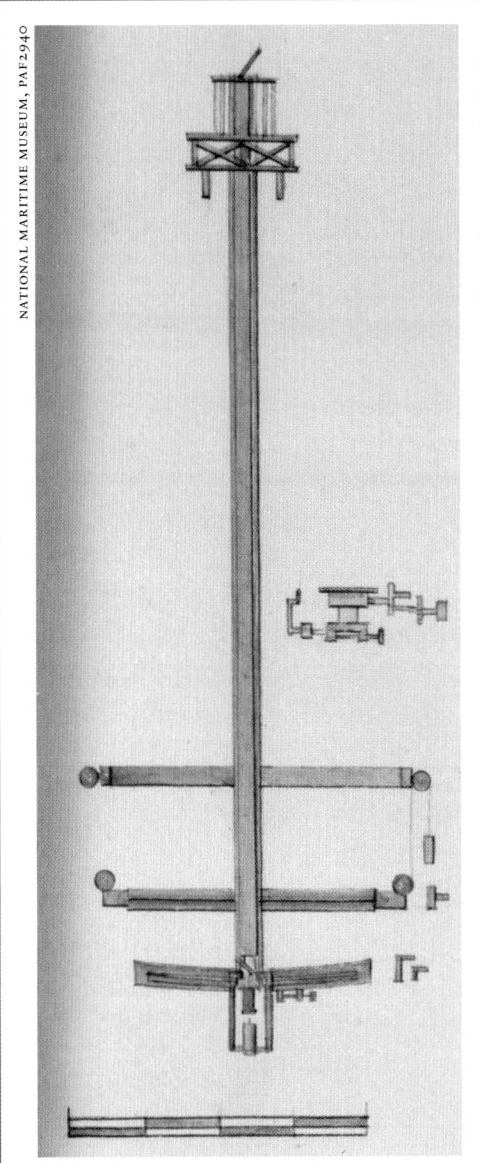

Fig. 2: This detail of a drawing by John Charnock, shows James Bradley's zenith sector at Greenwich in the late eighteenth century. Made by George Graham, it was long known as 'the famous zenith sector' because of its association with Bradley's discoveries.

Sirius was, with hindsight, never going to be possible with the available technology. However, he had set himself a number of other tasks. One, of course, was to try out Mayer's tables and the lunar-distance method of finding longitude during the voyages out and back. This proved the most important outcome of the expedition, alongside the opportunity it gave him to demonstrate his understanding of the use and adjustment of astronomical instruments. Also significant were his lengthy investigations into the cause of errors in his zenith sector.

Some of the tasks that Maskelyne took on were essential to the setting up of an observatory and to ensure reliable observations. These included assessing the going of the clock, checked against observations with the astronomical quadrant, and establishing the exact position of his observatory, using observations of the height of stars for latitude and of Jupiter's moons for longitude. He also made observations of the time and heights of tides and at least intended to record variations in the Earth's magnetic field, an investigation important to the use of compasses as well as having been considered a possible means of establishing position at sea.

One other significant area of research undertaken on St Helena (Fig. 3) was again inspired by the investigations of French astronomers. The problem under consideration was the shape of the Earth, which was thought to be an oblate spheroid after it was appreciated that the force of gravity varied with latitude, being greater at the poles than the equator. As the amount of time a pendulum takes to swing is inversely proportional to

Fig. 3: This 1815 hand-coloured engraving of the isolated Atlantic island of St Helena is after Thomas Sutherland (detail). Owned by the East India Company, this view of the island gives an impression of its strategic usefulness for trade vessels and its rugged volcanic terrain.

the square root of local gravity, it was possible to investigate the Earth's shape (the degree of its oblateness) by comparing the speed at which the pendulum of a clock swung at different latitudes. Thus Maskelyne had observed the rate of going of his clock and its pendulum at Greenwich, where it had been set up ahead of the voyage, and could now compare that with its rate on St Helena.

Information about the Earth's shape could help clarify that of its orbit and, once again, be fed into calculations of the Moon-Earth-Sun three-body problem. Maskelyne also approached the question of the shape of the Earth through observations of the Moon's parallax, which, again, would yield useful results if compared with similar observations undertaken in other latitudes.

After much work on St Helena and back in England on the errors of his zenith sector, Maskelyne was later to use one to see if he could detect the minute gravitational attraction exerted by a rocky mass on the plumb-bob incorporated into the sector. This phenomenon, predicted by Newton's theory, was later known as the Attraction of Mountains. In 1774, Maskelyne led a very successful expedition to observe it on the mountain of Schiehallion in Scotland, using the same zenith sector (with improvements that he felt had solved the problems that had previously dogged its use) and the same, or a very similar, clock.

In his 1775 paper to the Royal Society, Maskelyne declared his results, showing that the mountain exerted a measurable attraction on the plumb-bob; that this demonstrated the universality of gravitation that Newton had predicted, and that the Earth's mean density was at least double that at its surface (i.e. the density of the core was much greater than many assumed).[8] His demonstration of the effect of landmasses on the plumb-lines of astronomical instruments also had important implications for their use, and for assessing the reliability of observations that had not been corrected for such errors. This included the astronomical measurement of meridian arcs, such as that undertaken by Pierre Louis Maupertuis in 1736 in an attempt to resolve the question of the Earth's being oblate or prolate.

In all of these series of observations, we can see Maskelyne's full engagement with Enlightenment astronomy, its experimentation with instruments and its use of geographic distance to improve knowledge of the motions of heavenly bodies. It was knowledge that could be applied to navigation and survey work, and involved techniques that were used by the astronomers who, from the later eighteenth century, were regularly included in naval voyages of exploration. This was a business with which Maskelyne was to be intimately involved (as discussed further in Case study E), starting with the voyages of James Cook, the first of which was an expedition to Tahiti to observe the 1769 transit of Venus.

These experiences in expeditionary or field astronomy were, of course, not unimportant to Maskelyne's ability to claim expertise in managing a permanent astronomical observatory. In such institutions, the business of meridian astronomy – that is measurements made with fixed instruments aligned north-south, to map the stars and to measure the motions of the Sun, Moon and planets – was undertaken as an essential backdrop and support to these more global ventures, which in turn fed back information that could correct regular practice. It is to Maskelyne's management of one such establishment, the Royal Observatory at Greenwich, that the next chapter turns.

3

MASKELYNE THE MANAGER

Nicky Reeves

INTRODUCTION

Upon becoming Astronomer Royal in 1765, Nevil Maskelyne inherited two traditions: worrying about the instability and fragility of instruments, and worrying about how to manage and bring to publication large quantities of observations. The Royal Observatory at Greenwich was but one of many locations where delicate astronomical and horological instruments were used and observations made, and its instruments and techniques were not necessarily superior to those employed elsewhere. The transit of Venus of 1761 revealed a variety of practices that were found wanting. He noted publicly and forcefully how, for instruments and practices at Greenwich to be the most exacting, regulated and secure, it was necessary for 'constant, regular attendance' from the Astronomer Royal, who would promptly share and circulate observations and expertise for the benefit of the nation.

Maskelyne highlighted how continuously demonstrating the exactness of instruments was crucial for the Greenwich Observatory to become the 'best' in Britain. This chapter considers how, in the years before and immediately after he became Astronomer Royal, Maskelyne described, lobbied for and then implemented a set of managerial, publishing, and epistolary practices that would secure his vision of British state-sponsored astronomy co-ordinated from Greenwich.

A MULTITUDE OF OBSERVATORIES

The Royal Observatory and the position of Astronomer Royal were established and maintained with government funds. In this sense, the Greenwich site where Maskelyne took up residence in 1765 was unique: it was the only state-funded observatory in Britain, managed by the nation's only state-funded astronomer (Fig. 1).

This did not mean, however, that Greenwich was necessarily the best funded or most extensively equipped observatory in the country, or even in London, and neither did it mean that Greenwich necessarily had the best clocks, telescopes or mural quadrants.

For example, Britain's most expensive observatory – and probably the world's – was maintained from 1739 by the Earl of Macclesfield in the grounds of his private residence, Shirburn Castle, Oxfordshire. In many ways it was better equipped than Greenwich for meridian observations. At a meeting of the Royal Society in 1747, James Bradley, Astronomer Royal since 1742, lamented that Greenwich had only one large mural quadrant,

Fig. 1: A watercolour of the Royal Observatory, Greenwich, from the south east, painted by John Varley around 1830. The River Thames and London are visible on the right of the painting.

which, being 'too heavy to be conveniently removed; and the Room wherein it is placed, being too small to admit of its being turned to the opposite Side of the Wall', could only be used to observe 'Objects' lying southward of the zenith, and thus the stars nearest to the pole could not be observed at Greenwich 'with proper Exactness'. 'Until this Defect in the Apparatus belonging to the Royal Observatory be removed', Bradley noted, Greenwich would be forced to rely on Macclesfield's observations made at Shirburn of the more northerly stars made with his mural quadrant which, in contrast to that at Greenwich, could be erected on both sides of a wall in the more capacious and versatile Oxfordshire observatory.[1] Bradley was not the first Astronomer Royal to highlight the inadequacy of the instruments at Greenwich and he would not be the last.

Thirty years later, in 1777, by which time Bradley's lobbying had gained several new instruments including the desired additional quadrant, the Swedish astronomical tourist, Thomas Bugge, nevertheless described not Greenwich but Thomas Hornsby's observatory at Oxford as 'the best in Europe, both as regards the arrangement and the instruments'. He further noted that the private observatory of the wealthy merchant Alexander Aubert in Deptford, near Greenwich was 'the most complete in Europe for its size'.[2] There were, furthermore, three not inconsiderable observatories in Cambridge colleges, at Pembroke, Trinity and St John's, the last of which possessed a pendulum clock that was possibly more reliable than any at Greenwich.

In 1769 the Reverend William Ludlam published a volume of astronomical observations made in the observatory he had contrived on top of a tower overlooking the first court of St John's College. This work contained unusually detailed descriptions of his astronomical instruments, including his regulator clock. Comparing the extent to which this clock lost or gained time over several months with the extent to which Nevil Maskelyne's regulator at Greenwich did the same, Ludlam noted that 'it seems therefore, that the clock at St John's College has been more regular than that at Greenwich'.[3] Ludlam's skill in practical mechanics and astronomy was so valued by the Board of Longitude that, in

1765, he was one of the committee of five whom they appointed to examine John Harrison's famous timekeeper for determining the longitude at sea. Ludlam's regulator had been made by John Shelton, who, in 1749 had constructed, under the direction of the internationally celebrated clockmaker George Graham, the same clock at Greenwich that Ludlam had found wanting. Hence this was not a fanciful claim by a provincial and inexpert amateur with inferior instruments, but rather a reasonable suggestion that, in 1769, the most reliable astronomical clock in the country (and probably the world) was to be found in Cambridge, not Greenwich.

FLAWED OPERATIONS AT GREENWICH

Greenwich was, therefore, but one of many sites where astronomical observations were made, and it had no special claim to any form of functional superiority merely by virtue of being the Royal Observatory. It was just a building containing some fairly expensive instruments, and there were many other structures in similar use in London and elsewhere, some of them called observatories, some called colleges, some warehouses, houses, or workshops. For example, from the 1720s until his death in 1751, George Graham made frequent observations from the roof of his Fleet Street address, often subsequently publishing them. This rooftop became a meeting place for European savants, both to learn and to display their skills. The Swede, Anders Celsius, now better known for his temperature scale, joined Graham on 15 March 1736 to observe and precisely time an eclipse of the Moon using a reflecting telescope, probably made by James Short, with a magnifying power of sixty-three and, presumably, one of Graham's own regulator clocks. Short himself made frequent observations from the roof of his house and workshop in Surrey Street, off the Strand, and published about forty papers in the Royal Society's *Philosophical Transactions* between 1738 and his death in 1768, including accounts of observations of eclipses, comets, transits of Mercury and Venus, and improvements to lenses and pendulums.

Men like George Graham and James Short were thus expert

manufacturers and handlers of astronomical instruments who were also capable of making exacting astronomical observations and judging what made an astronomical observation exact. When, in his analysis of the multiple observations of the transit of Venus of 1761, Short declared there to have been a serious operational flaw in the manner in which the transit had been observed at Greenwich, his criticism was therefore taken seriously. Greenwich did not necessarily have the best instruments, it seemed, but it also did not have the best observational practices.

The Royal Society purchased a regulator clock by John Shelton for the St Helena mission that it organized to observe the 1761 transit of Venus there: it also bought one by John Ellicott for the parallel expedition to Bencoolen. Bradley's assistant, Charles Mason, was sent to Bencoolen with the surveyor Jeremiah Dixon, while the young and well-connected Cambridge graduate Nevil Maskelyne was sent to St Helena. Three reflecting telescopes 'executed' by James Short were made available for St Helena, Bencoolen and Greenwich. Short himself observed from Savile House, the Leicester Square residence of Edward, Duke of York, in the presence of the Duke (George III's brother and second in line to the throne), the king's three other brothers and their mother, Princess Augusta. Short used one of his own telescopes and an astronomical clock by Shelton, which Short described as 'the fellow of that which was last made for the Royal Observatory at Greenwich'.[4] The transit was observed at Greenwich by three people, using three different telescopes. Bradley was present, 'though not in a condition to observe because of his bad state of health'. Nathaniel Bliss, Savilian Professor of Geometry at Oxford (and soon Bradley's successor as Astronomer Royal, 1762–4), observed with a fifteen-foot refracting telescope that Bradley had brought to Greenwich in 1742, and for which he was paid £20 in 1748 when it became the property of the Observatory. Charles Green, the Astronomer Royal's new assistant, used a two-foot reflecting telescope lent by Short, and the instrument maker, John Bird, used an eighteen-inch reflector 'of his own making'.[5]

In two papers published in the *Philosophical Transactions*, Short collated and compared the observations of the timing of the moment that Venus appeared to make its second contact with the edge of the Sun made at fifteen European sites and at the Cape of Good Hope (the Bencoolen voyage was delayed and then diverted to the Cape, and cloudy weather scuppered the St Helena observations). He wrote that everywhere this observation had been made successfully

> and where there were more observers than one, we find a difference in the time of each observer; the observation at Greenwich is an exception to this, as the three observers all agreed to the same second, in the observation of the contact of Venus with the Sun's limb; which is the more surprising as they used telescopes of different constructions and of different magnifying powers.

Such exact agreement made it look suspiciously as though the independence of the three observers at Greenwich had been compromised. The coincidence astonished Short until he heard of the circumstances from Thomas Hornsby, when

> his surprise was at an end, for he had been informed at Greenwich, that Mr Green, the assistant observer there, as soon as he judged that the internal contact was formed, called out *now*. This must certainly have caused some disturbance to the other observers, and might possibly influence their judgement: and the fact (as I am informed) was each Observer had a second watch in his hand, and instantly stopped their watches, each having his hand at his watch ready to stop.

Short noted that the coincidence in timings at Greenwich of the external contact still needed to be accounted for because 'Mr Green did not call out at this time, because he was forbid by Dr Bradley': the problem was that the proximity of the three observers meant that 'the instant one of the observers stopped his watch, may it not be presumed that the noise of the nicking of it might be heard by the rest? especially as there was a profound silence during the time of the observation'.[6]

The observation at Greenwich of the most important

astronomical event of the century was poorly managed: Bradley, the Astronomer Royal, was too ill to observe, and the observational practices ill disciplined and ill defined, whereby the critical independent judgement of individual astronomers of the exact moment at which Venus made contact with the Sun was compromised. Ironically, the remote and temporary observatory at the Cape was easier to manage than Greenwich: Mason and Dixon each observed the transit with the two-foot reflecting telescopes provided by Short, timed with a regulator by Ellicott. The four seconds of time difference between these two observations, unlike at Greenwich, was noted approvingly by Short. A temporary wooden observatory erected at the southern tip of Africa with the possible assistance of unnamed labourers provided by the Dutch authorities was, on this occasion, a more effectively managed site than the Observatory that had existed at Greenwich for eighty-six years.

CLOCK WORK

Short's criticisms were not limited to telescopic observation. He also got involved in discussions about the gravitational experiments with pendulum clocks carried out at the Cape and St Helena. Following the observation of the transit at the Cape, Mason sailed with the Ellicott clock to St Helena, where he joined Maskelyne, who had already made detailed observations of the going of his Shelton clock both at Greenwich and on the island. Together, they now did the same thing for Mason's Ellicott clock, comparing its going at the Cape and at St Helena. Maskelyne recognized that 'the exactness of the going of a clock depends upon the firm manner of setting it up', and in his later report to the Royal Society emphasized how he securely fastened the Shelton clock to a stone wall with large pieces of wood and screws, the heavy clock case resting upon a large flat stone: Mason secured the Ellicott clock in a similar manner.[7]

Maskelyne and Mason were certain that these precautions ensured the stability of the clock such that the relative strengths of gravity at St Helena and Greenwich could be compared. James Short strongly disagreed. As he wrote in 1762:

> No observations of the difference in the going of a clock, made at different places, can, with certainty, determine the difference of the effect of gravity at these places; because it has been found, by experience, that the same clock, placed at different times on different walls, in the same room, will make a difference in the going of the clock; even though every part of the clock remains the same.[8]

Maskelyne replied in 1764:

> I readily allow that, if clocks are fixed up in a slight manner, or against common wainscots, the experiments made with them cannot be depended upon. Nevertheless it does not appear, but that when they are fixed in a firmer manner, they may be depended upon near enough to be of considerable use in physical enquiries.[9]

Given that, in late 1764, both Short and Maskelyne were considered serious candidates for the then-vacant position of Astronomer Royal, these were timely and fundamental debates about the correct management of instruments. Maskelyne emphasized the need to demonstrate at all times the effective management and maintenance of clocks and other instruments. This was his response both to Short's criticism of managerial practices at Greenwich and to Short's scepticism as to the value and certainty of pendulum clocks. Managerial oversight would secure instruments and so secure the value of future observations.

Greenwich in 1765 possessed three main pendulum clocks, all from the workshop of George Graham. The first two, referred to by astronomers and historians alike since around 1840 as 'Graham 1' and 'Graham 2', were acquired in 1725 by the second Astronomer Royal, Edmond Halley, for £12 each. The third clock, 'Graham 3', the one found wanting by Ludlam in 1769, was acquired by his successor, Bradley, around 1750 (Fig. 2).

NATIONAL MARITIME MUSEUM, ZBA0709

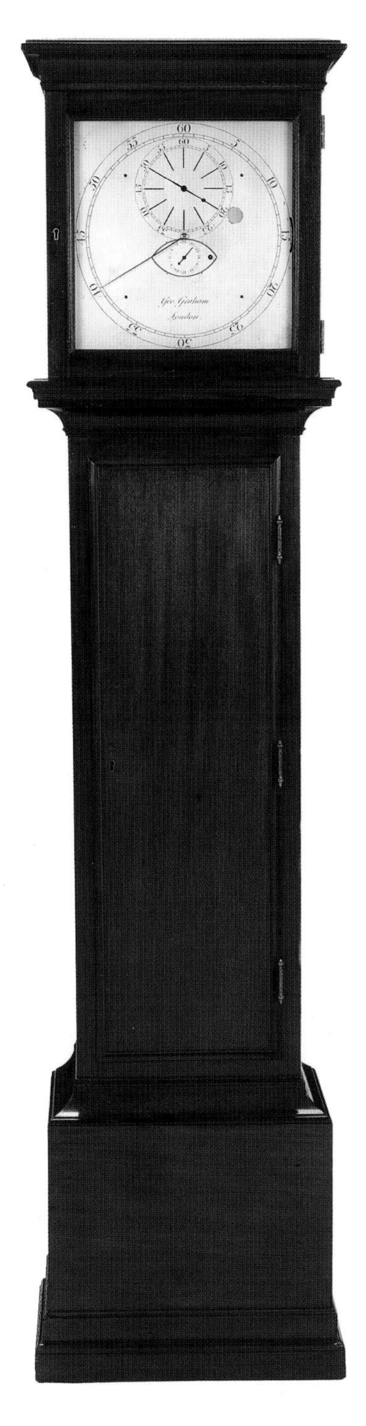

Fig. 2: The astronomical regulator known as Graham 3, made by George Graham in about 1750 for James Bradley, the third Astronomer Royal.

It cost £39 and again was provided by George Graham, having been made under his supervision by John Shelton. These clocks were used to determine the precise timing, or right ascension, of the daily transit over the meridian of the Moon, the Sun, planets and certain prominent bright stars. These observations were made with a transit instrument, an eight-foot telescope fixed in the meridian and movable about a horizontal axis (Fig. 3).

When Maskelyne arrived in post, the transit instrument was that contrived for Bradley by John Bird in 1749 at a cost of £73 16s 6d, and it was this instrument that defined the Greenwich meridian from the early 1750s until 1816. Neither Bradley nor Maskelyne thought it the function of the Observatory to produce a complete map of the heavens. Rather, what was needed was a highly selective one of the positions of what Maskelyne called '34 principal fixt stars', measured in terms of right ascension. These were stars that were bright and therefore 'more frequently visible

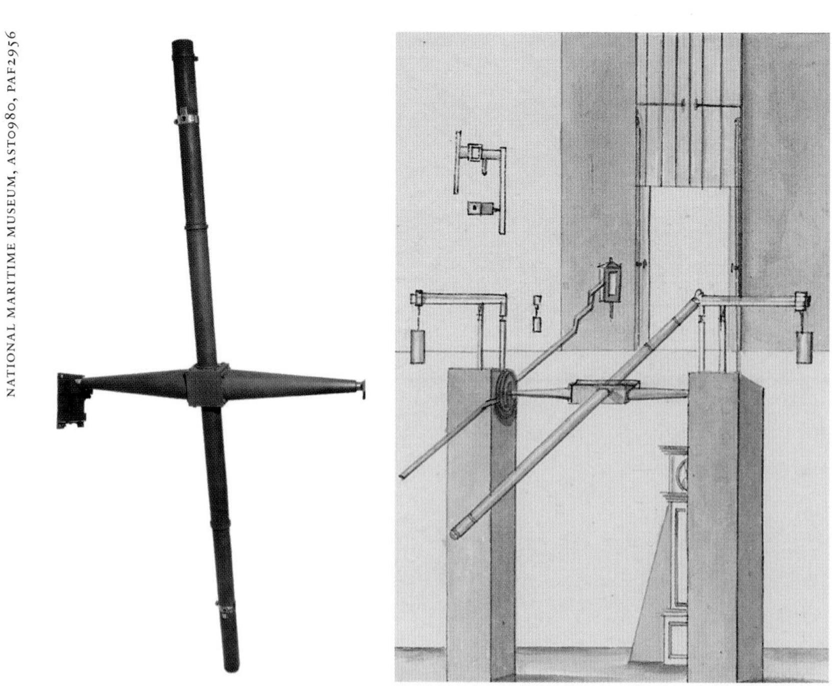

NATIONAL MARITIME MUSEUM, AST0980, PAF2956

Fig. 3: The Royal Observatory's transit instrument, made by John Bird in about 1749. The pen and wash drawing, by John Charnock, shows the same instrument in situ at the Observatory, in the late eighteenth century.

in the day time than smaller stars, and not so easily obscured by haziness of the air or thin clouds, either by day or night'. They were also near to the equator and hence 'they pass the vertical wires of the transit instrument with greater velocity than stars remote from the equator, and consequently the times of their transits may be more exactly observed'.[10]

As Maskelyne later described it in 1776:

> the observations made with the transit instrument are the most numerous of any day; besides the transits of the planets, the transits of several fixed stars being observed every day that the weather permits, to be compared with those of the planets, in order to settle their right ascensions, and serving also to determine the clock's rate of going from day to day.[11]

The frequency and regularity of these observations was invoked by Maskelyne to distinguish Greenwich from the many other competing public and private spaces in which meridian observations were made. Furthermore, in any location, the delicacy and fragility of instruments, and the buildings, rooms and walls that housed them, meant that their value and precision had to be repeatedly demonstrated. Clocks, and all other major instruments in the Observatory, were used by Maskelyne not only to *make* precise observations but also to *demonstrate* that precision. In this he was returning to an old problem. For example, one of the earliest tasks for the first Astronomer Royal, John Flamsteed, in the 1670s, was to determine through astronomical means whether or not the Earth rotates on its axis at a constant (isochronic) speed, a critical question for the very possibility of accurately mapping the heavens. To do this, he had to demonstrate his continuing control over his instruments.

In July 1677 Flamsteed told Richard Towneley, a Lancashire associate of Flamsteed's patron Jonas Moore, that

> our clocks kept so good correspondence with the heavens that I doubted not that they would prove the revolutions of the Earth to be isochronical ... I can now make it out by three months

> continued observations, though to prove it fully it is requisite the clocks be permitted to go a whole year without any alterations.

Flamsteed hailed 'the wonderful goodness of the regulators' while constantly noting their actual irregularity. Clocks slowed down in an unpredictable way, or stopped altogether, through being affected by strong winds, changes in humidity, or becoming clogged with dust and dirt.[12] Their constant maintenance, as well as detailed reporting of their behaviour, was necessary for a well-oiled observatory. Demonstrating that a clock could 'go a whole year without any alterations' was equally central in the 1760s to Maskelyne's implied model for the proper management of an observatory, and he took advantage of his two expeditions to remote sites in 1761 and 1764 to demonstrate to the Royal Society, the Board of Longitude and the wider astronomical community just how well he could manage fragile clocks and large instruments in general.

As discussed in Case study B, Maskelyne used the opportunity of his stay on St Helena in 1761 to undertake detailed experiments with clocks, as well as attempting to determine the annual parallax of the star, Sirius, using a specially contrived ten-foot telescope called a zenith sector. While he failed to observe the parallax, he managed to turn the observational failures into managerial successes. He undertook an unusually long investigation into the material instability of his temporary observatory, clocks and telescope. A crucial component for the 'exactness' of the zenith sector was its plumb-line and plumb-bob, which defined the zenith, and this delicate suspended mass, just like the pendulum of a pendulum clock, required exacting and skilful care. At the Royal Society, and in a theatrical demonstration in the Great Hall at the British Museum, he explained how he had set up his pendulum clock and zenith sector and then proceeded to observe over many months how ambient conditions, creaking floorboards, his own motion in and around the temporary observatory, and the effects of actually using the instruments all affected their exactness, but could be controlled and accounted for through careful managerial attention.

One of the first things Maskelyne did upon becoming Astronomer Royal in 1765 was to declare many of the instruments and clocks at Greenwich unsatisfactory. He claimed, for instance, that his zenith-sector experiments on St Helena had revealed a hitherto unrecognized flaw in the manner in which all previous plumb-lines on sectors and mural quadrants had been hung. In collaboration with the instrument maker, John Bird, he asserted that without alterations to plumb-line suspensions 'the principal observations [at Greenwich] must be at a standstill', thus using his experiences at St Helena to underline his expertise in managing instruments and, implicitly, to criticize his immediate predecessor, Bliss, for allowing those at Greenwich to get into disrepair.[13]

Travelling to Barbados in 1764 on behalf of the Board of Longitude, Maskelyne was given the task of determining the island's longitude through astronomical observations of the moons of Jupiter. This was to assist adjudication of the ability of John Harrison's timekeeper to keep time with sufficient exactness to be relied upon to determine longitude after a transatlantic voyage. While in Barbados, Maskelyne also took the opportunity to demonstrate just how a pendulum clock could 'go a whole year without any alterations', or at least go for about ten months. Just as with his St Helena experiments, he kept a detailed daily record of 'vibrations' and 'wobbles' of the pendulum clock, noting who touched it and who had access to it, including servants and assistants, how it was 'fixed' and 'fastened', and how all of this affected its going. Implicit here was the idea that for an observatory to secure instruments and observations, it needed to be constantly attended to, and that this attendance itself needed to be documented in detail and shared in manuscript and in print. It was this constant attendance, rather than, for instance, superior instruments, that from the mid-1760s onwards distinguished Greenwich from the many other sites where astronomical observations were made. In his lobbying of the Royal Society in the early to mid-1760s, Maskelyne fiercely criticized the management of instruments and observations at Greenwich, and argued that the 'due execution of the office of Astronomer Royal' demanded a constantly attending and regularly and promptly communicating custodian residing at Flamsteed House. Maskelyne certainly showed excellent timing in taking advantage of the fact that the Observatory had been neglected since 1758 and used this recent unsatisfactory history to lobby for his particular view of how it should be managed.

CONSTANT ATTENDANCE

The lunar-distance method of determining longitude at sea required the Moon's future position to be calculated and, perhaps, distributed in print. By mid-century European astronomers, such as Nicolas Louis de Lacaille and Tobias Mayer, appreciated that a theoretical and mathematical account of the Moon's motion that could yield predictions of its future position needed to be allied with, and corrected by, detailed lunar observations made with a transit instrument, mural quadrant and pendulum clock. In 1755, Mayer sent a manuscript of his newly calculated lunar tables to Lord Anson, First Lord of the Admiralty and Commissioner of the Board of Longitude. Bradley, the Astronomer Royal, reported in 1756 that he had compared positions of the Moon calculated by Mayer with over 230 observations that he had himself made at Greenwich with the transit instrument and eight-foot mural quadrant (Fig. 4), noting that he 'did not find any difference so great as 1'½ between the observed longitude of the moon and that which I computed by the tables'.[14]

In April 1760, Bradley reported that over the next four years he had made more than 1,100 further lunar observations from Greenwich that confirmed the close fit between the observed and computed positions of the Moon, indicating the usefulness of Mayer's tables for determining longitude at sea. It was these that Maskelyne worked with on his voyage to St Helena, in producing his 1763 *British Mariner's Guide* – which also included instructions and worked examples on how to use a Hadley's quadrant (known later as an octant) – and on the later trial of the lunar-distance method on a voyage to Barbados in 1763–4.

NATIONAL MARITIME MUSEUM, AST0971

Fig. 4: The eight-foot mural quadrant at the Royal Observatory, made for James Bradley by John Bird in 1750.

In addition to this publication, in June 1763 Maskelyne told the Royal Society that if he had been given access to Bradley's observations of Sirius made at Greenwich in the 1750s, he would have known that a determination of an annual parallax of that star was simply not possible given the limits of precision or 'exactness' of his instruments, but Bradley 'always declined giving me any sight of them'. Regarding observations of the Moon made at Greenwich throughout the 1750s, he also informed the Society that

> the late Mr Mayer, a learned foreigner has been enabled to run away with the creditof bringing [lunar theory] to a great degree of perfection, by having a set of observations of the moon, made at the Royal Observatory communicated to him in particular; which, had they been as freely imparted to the learned of our own nation, would have preserved the honour intire to ourselves.

Bradley had shared observations made at Greenwich with Mayer but not with other, British, astronomers. 'Thus, by an unhappy concealment and retention of the observations made at the Royal Observatory, the improvements of astronomy have been kept 50 years behind hand', Maskelyne declared. He emphasized how important it was not just to make observations but also to share them. Past Astronomers Royal, he said, had been too possessive of manuscript observations, and too quick to treat them as private property, Bradley being merely the most recent. 'Mr Flamsteed the first Astronomer Royal's tenaciousness in withholding his observations, not only from the public but even from his learned friends, a long while deprived Newton of the means of establishing his theory of the moon', Maskelyne told the Society, and 'with respect to Dr Halley's Observations, made with superior instruments to those of Mr Flamsteed we have been even more unfortunate; being entirely deprived of them, his executors having seized and to this very day detaining them as their own property'.[15]

Bradley had died in July 1762, with his twenty years of observations at Greenwich unpublished. In June 1763 Maskelyne asked the Royal Society:

> respecting the late Dr Bradley's Observations, what loss would not the public sustain? What detriment would not astronomy suffer? What materials would not physics and the lunar theory lose? And what discredit might not fall even on this Society, if we should any longer allow these valuable and useful observations to be withheld and concealed, not only from the public, but also from the most inquisitive and able mathematicians?

This is exactly what did happen, when it became apparent in September 1764 that Bradley's complete observations had been removed by his executors, who had arrived at the Observatory to demand them from Nathaniel Bliss's widow, Bliss himself having died earlier that month after only two years as Astronomer Royal. Thirteen folio volumes of unpublished observations by Bradley had been removed and, despite Maskelyne's repeated attempts, remained unpublished and inaccessible until 1798.

It was in this light that the Royal Society, in its role as Visitors to the Observatory, drew up nine regulations in November 1764 for 'the due execution of the office of Astronomer Royal'. These demanded the regular publication of observations, the permanent residence of the Astronomer Royal at Greenwich, and the vigilant control of access to the rooms, instruments and manuscript observations. The first regulation noted that

> whereas the end of the institution and continuation of this office requires a constant, regular Attendance of the Astronomer Royal at the Observatory: it will be required that he do bona fide reside there, and that he do not accept of any office or employment the duty whereof to be personally performed by him, is incompatible with such bona fide residence, and that in case of absence above a few days from the said Observatory, the reason of such absence, shall be previously allowed by those who are or shall hereafter be appointed Visitors of the Royal Observatory.

The second demanded 'that the Astronomer Royal and his assistant shall never be absent from the Observatory both at the same time', and the third

> that no persons be admitted into the rooms of the Observatory where the instruments are kept, unless the Astronomer Royal or his Assistant shall be present, who are to take special care that no person be allowed to touch or handle the Instruments, so as to endanger the spoiling thereof, or doing any damage thereto.

The seventh regulation stipulated that the manuscript observations never be removed from the Observatory 'upon any pretence whatsoever', and the eighth required that 'true and fair copys [*sic*] of the annual observations made by the Astronomer Royal and his assistant shall be transmitted within six months after every year shall be elapsed to the Council of the Royal Society'.[16]

It was not until Maskelyne took up residence at Greenwich, ninety years after the Observatory's foundation, that there was any regular sharing of observations. Maskelyne shared them annually with the Visitors from 1767, in manuscript form, and they were later printed and widely circulated in the 1776 *Greenwich Observations.* Echoing his 1763 address, he noted that for 'the better attainment' of determining longitude at sea 'and for the general improvement of astronomy', 'voluminous tables of figures' of observations made at Greenwich needed to be published regularly.[17] The labour involved required almost constant attendance by either the Astronomer Royal or his assistant, in conformance with the nine regulations of 1764. Inspecting either the manuscript or printed observations (and the remarkable agreement between manuscript and print was in itself an extraordinary achievement: see Fig. 5) it can be seen that either he or (more often) his assistant was present in the transit room almost hourly from mid-1765 onwards.

Constant attendance meant constant control of instruments and manuscript observations, and hence a continuous demonstration of the exactness of the instruments. Maskelyne performed all of this in one of his first tasks after becoming Astronomer Royal, namely the ten months he spent at Greenwich from May 1766, 'trying' John Harrison's H4 timekeeper on behalf of the Board of Longitude, to determine 'the manner of its keeping time'. The crucial question was whether Harrison's watch could be used

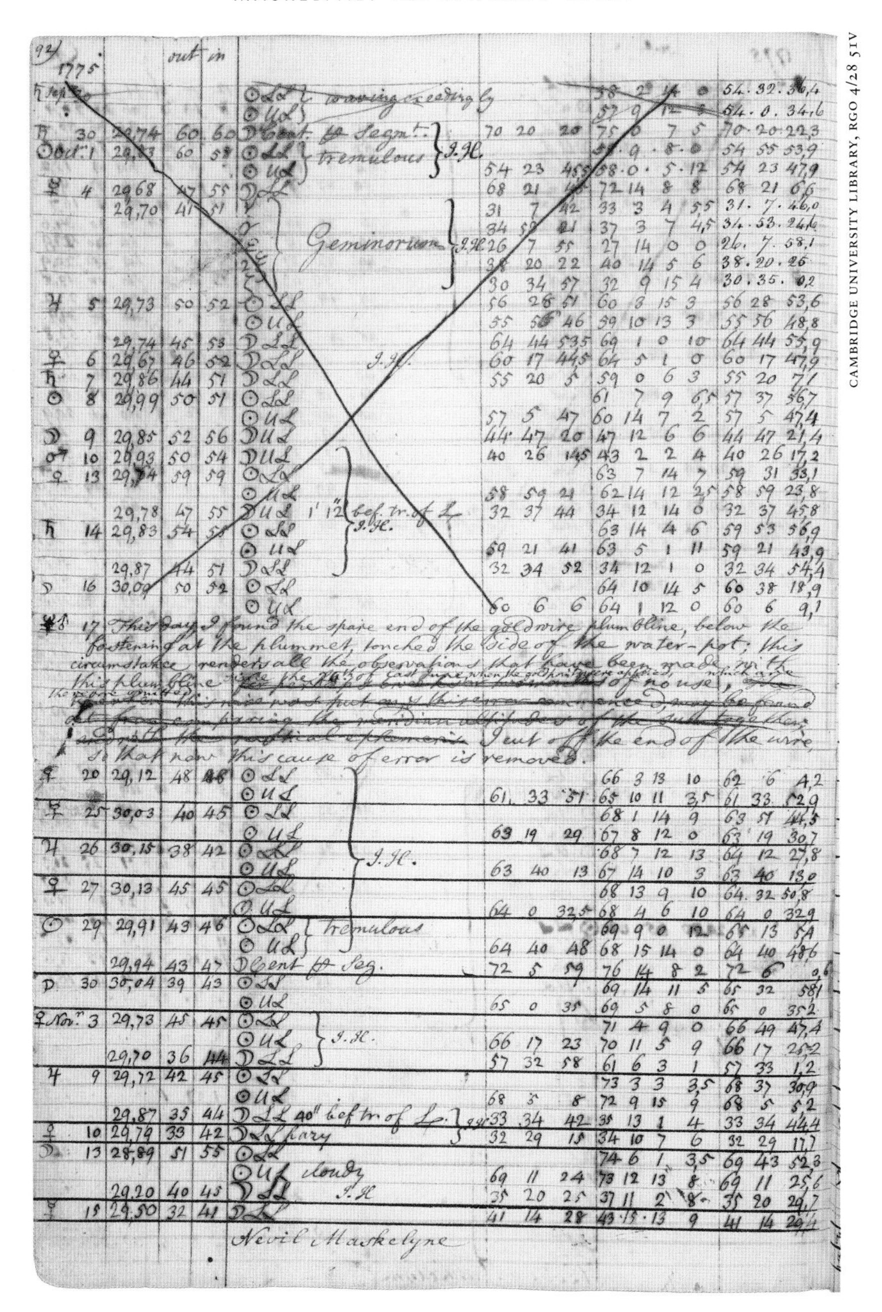
This day I found the spare end of the goldwire plumbline, below the fastening at the plummet, touched the side of the water-pot; this circumstance renders all the observations that have been made with this plumbline ... of no use; ... I cut off the end of the wire so that now this cause of error is removed.

Nevil Maskelyne

Fig. 5: Maskelyne's note in the centre of this page of manuscript observations, describing a problem with the plumb-line, and consequent need to omit the foregoing observations, was printed verbatim in *Greenwich Observations*.

to determine mean time over an extended period and thus be a potentially useful method of determining longitude at sea. All eighteenth-century clocks lost or gained time; determining the 'rate' or 'going' of a clock allowed for corrections to be made, but only if it could be demonstrated to have a regular rate of going.

In May 1766 Harrison's watch was delivered to Maskelyne, at the Admiralty, and then transported to the Royal Observatory accompanied by three witnesses, Captain Thomas Baillie of the Royal Hospital for Seamen at Greenwich, John Ibbotson, Secretary of the Board of Longitude, and watchmaker Larcum Kendall. The watch was secured in a box with two locks, each of which required two keys. The keys to one lock were held by Maskelyne, the keys to the other lock were entrusted to Baillie. For ten months, at a stated hour each day, an officer of Greenwich Hospital attended to assist in the unlocking of the box. The watch was then compared with the 'Graham 3' pendulum clock, either by Maskelyne or his assistant, and 'always done in the presence of, and attested by one of the officers of Greenwich Hospital'. The going of Graham 3 itself was 'deduced from the observed transits of the fixed stars over the Meridian ... whereby it appears to have kept very regular time': it was, as Maskelyne put it, 'compared continually'. A lengthy account of ten months' worth of daily comparisons of the watch and the pendulum clock was published 'by order of the Commissioners of Longitude' in 1767, including quantified barometric and thermometric readings, detailed descriptions of the manner in which the watch was wound up, by whom, and in what position it was left, frequent descriptions of the weather (and in particular the strength of the wind), and on one occasion an account of an alteration to the length of the pendulum of the clock.

Maskelyne's principal conclusion from this trial was that 'the rate of the going of the watch appears ... to be too irregular'. It did not go in a regular fashion, neither did the variation of its going correspond directly to variations in temperature, nor in any accountable way to the various positions in which it was placed, either horizontally, vertically or at any other inclinations. Furthermore, it seemed to Maskelyne 'that the watch cannot be taken to pieces and put together again without altering its rate of going considerably'.[18]

It was far from obvious that Maskelyne, who had never made a clock, was the most appropriate person to test the timekeeper, just as it was far from obvious that the Greenwich Observatory was the best place to conduct a land-based trial of it. The London workshops of several prominent clockmakers, for instance, might have been considered more appropriate sites for the trial. In overseeing the trial, Maskelyne was implicitly insisting that the expert judging of timekeepers was best managed by an astronomer, and emphasizing that it was the Astronomer Royal who was best placed to make judgements on timekeepers for the Board of Longitude. In later decades the rating of chronometers for the Board did, of course, become one of the key roles of the Royal Observatory. Most of all, however, he took the opportunity to demonstrate just how effective a highly regulated, controlled and managed observatory regimen could be, with access to instruments almost militarily controlled, and the detailed observations of the watch and pendulum clock promptly printed and circulated.

At the centre of it all was the pendulum regulator and it is interesting to reflect that Maskelyne's first published account of observations at Greenwich was not of stars or the Moon but rather a series of observations of a watch and a pendulum clock. Remarkably, the 1767 account of the going of Harrison's watch was the first detailed, published account by any Astronomer Royal of the daily behaviour of a pendulum clock there, Flamsteed's letters of almost a century earlier having lacked any quantification and being very much private communications. Maskelyne claimed that Graham 3 was 'very regular'. William Ludlam in Cambridge, as we saw earlier, thought otherwise: but it was only because of Maskelyne's readiness to communicate and circulate observations that Ludlam was able to make such a judgement. Under Maskelyne's command, Greenwich became a crucial node in a republic of (astronomical) letters.

AN EPISTOLARY CENTRE

From 1765, in contrast to previous customs, it was a novelty for an Astronomer Royal to respond promptly and reliably to correspondence, to share observational data cordially and pragmatically, and to circulate practical advice on observing techniques. Prior to Maskelyne's appointment, not a single item of correspondence on an astronomical topic, directed to the Astronomer Royal, had been published in the *Philosophical Transactions* since 1758. Numerous observations and enquiries from Britain and overseas on the 1761 transit of Venus were addressed to James Short, to Lord Macclesfield (the Society's astronomer President), to Charles Cavendish (the Vice-President), to the Society Secretaries, Charles Morton and Thomas Birch, and others, but none were addressed to Bradley. Not a single letter to Bliss, Bradley's successor, on any topic, appeared in the *Transactions* during his brief tenure.

The majority of accounts of the transit of 1761 lacked exactness. As lamented by no less than Jerome Lalande, Parisian astronomer and editor of the *Connaissance des Temps*, and communicated by Maskelyne to the Royal Society in November 1762, even the longitudinal separation of the Greenwich and the Paris observatories was 'uncertain'.[19] Consequently, any derived estimations of the solar parallax were also inexact. James Short had collated observations of the transit of Venus at fifteen different European sites, comparing them with Mason's and Dixon's observations at the Cape, as did Thomas Hornsby at Oxford. Both men shared their significantly different results with the Royal Society, agreeing that no measure of certainty had been achieved. Global correspondents such as John Winthrop at Cambridge, Massachusetts, understood this, and it was to Short that Winthrop wrote in 1764, providing further 'substance' regarding both the longitude of his temporary observatory at St John's, Newfoundland, and the timing of his observations of the transit three years earlier. Winthrop thanked Short for previous 'candid and judicious remarks', pointedly noting that he had sent an earlier account to Bradley, 'which I believe never reached his hands', and likewise that he had sent another in 1763 to Bradley's successor but 'had no return from Mr Bliss'.[20]

At the same time that Winthrop was writing to Short, accounts of meteors, and, in particular, the eclipse of the Sun of 1 April 1764, were addressed to the secretaries and various fellows of the Royal Society, often including Short, but never to the Astronomer Royal. Short reported that he had observed the eclipse from his house in Surrey Street, in the company of the Earl of Morton (who succeeded Macclesfield as President) and noted that they had both observed, 'but in different rooms, out of sight and hearing of one another.'[21] From Liverpool, James Ferguson wrote to Thomas Birch that he had calculated the estimated time of the eclipse in advance using 'Meyer's [*sic*] tables, as we have them published by Mr Maskelyne', making use of Maskelyne's newly published *British Mariner's Guide*, and adding that while 'several people came into the room to see the eclipse ... I told them before-hand that they must neither speak nor move till the eclipse was found to be begun'.[22] The emphasis expressed earlier by both Maskelyne and Short on managing and policing observatories, and circulating clear instructions, was patently having an effect, despite the lack of direction from Greenwich itself. Somewhat unfortunately, by contrast, the Astronomer Royal, Bliss, reported that he had made the journey from Oxford, where he resided, in order to observe the eclipse at Greenwich, but a 'watery defluxion on my eyes occasioned by a cold ... obliged [me] to wipe my eye perhaps at the very time of the contact'.[23]

By the time of the 1769 transit of Venus, Maskelyne was, in contrast, already engaging with numerous correspondents, requesting, receiving and interrogating their observations, having previously published detailed practical instructions on how to observe the transit. Correspondents round the world thanked him for the advice and instructions found in his 1763 *British Mariner's Guide*, his widely circulated 1768 *Instructions Relative to the Observation of the Ensuing Transit of the Planet Venus*, and numerous epistolary communications. Maskelyne, like Short and unlike Bliss, clearly read and answered his mail. The 1768

Instructions went into great detail regarding setting up telescopes and clocks, observational techniques and the correct way (according to Maskelyne) of recording observations. Astronomers sent by the Royal Society to Hudson's Bay, the North Cape (Norway) and the South Pacific, received detailed instructions; a temporary wooden observatory for the Hudson's Bay party was prefabricated in England and tested by Maskelyne at Greenwich. He then entered into extended correspondence with observers in order to extract from them the most detailed information possible. For example, from Hawkhill near Edinburgh, the physician, James Lind, sent observations of the transit by himself and two associates. Maskelyne replied promptly, initiating a lengthy exchange in which he pressed Lind for details on instruments used, the longitudinal distance between Hawkhill and Edinburgh Castle, and the individual circumstances of each of the three observers' perceived times of contact.[24] From Massachusetts, John Winthrop, who had previously 'had no return from Mr Bliss', wrote to Maskelyne, who presented the observations to the Royal Society. Observations were received from correspondents as far afield as Quebec, Gibraltar, Stockholm and East India Company engineers stationed in Dinapoor, all of which invariably expressed gratitude for Maskelyne's helpful *Instructions*.

Greenwich in the 1760s was but one of several places that possessed and made use of clocks, quadrants and telescopes by London's most celebrated instrument makers. Maskelyne claimed that astronomical and horological precision, or as he called it 'exactness', was to be achieved through the practices of discipline and vigilance. The Astronomer Royal needed to be a fierce gatekeeper, controlling access to the instruments and recognizing how manuscript observations must be protected at all times. At the same time, his role required him to behave as what he also was, a civil servant of the State, cordially and promptly communicating expertise, instructions and observations.

Under his predecessors the Royal Observatory had witnessed

many of these things, but Maskelyne emphasized that, at critical points in its recent history, security and attention had been lacking, resulting in unsatisfactory observations and the catastrophic loss of manuscript records. Maskelyne's managerial attitude extended beyond the Observatory, of course: as the following chapter shows, for instance, the disciplining of employees and human computers was crucial to the success of ongoing projects like the *Nautical Almanac*. While his appointment in 1765 did not instantly make the Greenwich Observatory the country's dominant astronomical and horological institution, it was through his insistence on managerial reform that it attained this status by the end of the century.

FURTHER READING

By far the best way to get a feel for the international correspondence network centred on the Royal Society's premises in Crane Court, off Fleet Street, is to read the Society's printed journal of record, the *Philosophical Transactions*, which is freely available online at http://rstl.royalsocietypublishing.org/content/by/year. Remarkably, Charles Weld's *A History of the Royal Society, with Memoirs of the Presidents* (London: Parker, 1848) remains the best book-length account of the Royal Society in the eighteenth century. Harry Woolf's *The Transits of Venus: A Study in 18th-Century Science* (Princeton: Princeton University Press, 1959) unsurprisingly covers the history of astronomy in the 1760s in great detail but also serves as a model for understanding the 'prolonged liaison between science and government' that constituted the numerous and unprecedentedly global transit of Venus projects undertaken by many nations.

Several of the instruments and clocks mentioned here survive in various collections. The Shelton regulator which William Ludlam in 1769 declared superior to Maskelyne's at Greenwich is on display and in working order in the library of St John's College, Cambridge. The Shelton regulator now in the National Museum of Scotland in Edinburgh is the best candidate for being the one purchased by the Royal Society in 1760 for the St Helena mission. Maskelyne took this clock to Barbados in 1763. It was later sent by him to Mason and Dixon in America to aid their survey of the boundary of Pennsylvania and Maryland, and to undertake gravitational experiments. It then went with James Cook and Charles Green to Tahiti for the 1769 transit of Venus, and was used by Maskelyne himself in 1774 on the Scottish mountain, Schiehallion, to undertake experiments into the density of the Earth. The regulator known as Graham 3 is in the collection of the National Maritime Museum, Greenwich (ZBA0709) and on display at the Royal Observatory. The zenith sector that Maskelyne took to St Helena in 1761 was last seen in the Royal Society's warehouse in 1789 when the instrument maker Jesse Ramsden inspected it: a similar device, with an improved method of suspending the plumb-line, was made by John Bird for the Radcliffe Observatory at Oxford in the early 1770s and is today on display at the Museum of the History of Science, Oxford. The Harrison timekeeper, H4, is today also on display at Greenwich, behind glass and secured by several locks, just as it was in the 1760s.

Bradley's relationship with Tobias Mayer is discussed in Chapter 6 of Eric Forbes's *Tobias Mayer (1723–62): Pioneer of Enlightened Science in Germany* (Göttingen: Vandenhoeck and Ruprecht, 1980). The new regulations for the conduct of the Astronomer Royal and the management of the Greenwich Observatory are discussed in P.S. Laurie's article 'The Board of Visitors of the Royal Observatory 1: 1710–1830', from the *Quarterly Journal of the Royal Astronomical Society*, 7 (1966), 169–85 and my '"To Demonstrate the Exactness of the Instrument": Mountainside Trials of Precision in Scotland, 1774', *Science in Context*, 22 (2009), 323–40, the latter of which also discusses Maskelyne's experiments on St Helena in 1761. The question of the purpose and scope of meridian astronomy is discussed in David Dewhurst's 'Meridian Astronomy in the Private and University Observatories of the United Kingdom: Rise and Fall', *Vistas in Astronomy*, 28 (1985), 147–58, and Jim Bennett's 'The English Quadrant in Europe: Instruments and the

Growth of Consensus in Practical Astronomy', *Journal for the History of Astronomy*, 23 (1992), 1–14.

Flamsteed's investigations to demonstrate the regularity of a clock are discussed in Derek Howse's 'The Tompion Clocks at Greenwich and the Deadbeat Escapement, Part 1 — 1675–1678', *Antiquarian Horology*, 7 (1970), 18–34. The global history and travel of Shelton regulators is discussed in Derek Howse and Beresford Hutchinson's 'The Saga of the Shelton Clocks', *Antiquarian Horology*, 6 (1969), 281–98. Maskelyne's trial of Harrison's timekeeper at Greenwich is discussed in Jim Bennett's 'The Travels and Trials of Mr Harrison's Timekeeper' in *Instruments, Travel and Science: Itineraries of Precision from the 17th to 20th Century*, eds Marie Noëlle Bourguet, Christian Licoppe and H. Otto Sibum (London: Routledge, 2002).

References to Nevil Maskelyne's global epistolary network into the 1770s and beyond are too numerous to name individually: one nice recent example, which shows his reach in British North America, is Stephen J. Hornsby's *Surveyors of Empire: Samuel Holland, J.F.W. Des Barres, and the Making of the Atlantic Neptune* (Montreal and Kingston: McGill-Queen's University Press, 2011).

CASE STUDY C

THE ASTRONOMER ROYAL AT GREENWICH

The institution and home that Maskelyne entered in 1765 was inherited from his predecessors. While he was clear about the changes he was making to the working regime of the Royal Observatory in terms of constant superintendence and distribution of its observational results, the instruments and existing fabric of the buildings were largely retained if, in various ways, altered and improved. Maskelyne was by and large happy to keep instruments that had been in use since Bradley's time, although in 1792 he did request a mural circle to replace the two quadrants, when he noted the extent to which these (British-made) instruments were now used in European observatories. However, it was not until 1806 that he received permission to order one, to replace what was by then confirmed as an inaccurate quadrant.

In the meantime he did add to the smaller instruments of the Observatory and improved others by incorporating the new achromatic lenses by George Dollond (which reduced the problem of chromatic aberration, caused by refraction of different wavelengths of light). An interesting account of Maskelyne's apparently rather experimental – or improvisational – approach to the improvement of his instruments comes from a description of the Observatory in 1777 by the Danish astronomical tourist, Thomas Bugge. He described how Maskelyne had attempted 'to eliminate the influence of the solar rays on the tube' of his transit instrument by 'covering it with white paper'. Bugge was 'more inclined to think that this would have a bad effect and increase the change of the tube by heating', since the paper was 'pasted immediately on to the brass' rather than being used as a screen.[1] Elsewhere, Maskelyne had constructed devices for screening observations from light and a plumb spirit-level.

The observing rooms, either side of the assistant's apartments, were also inherited spaces, although they were altered in a number of ways to improve observing conditions. Together they made up the New Observatory that Bradley had managed to persuade the Board of Ordnance to build when he arrived in Greenwich in the 1740s. The meridian wall, on which the two mural quadrants were fixed, in fact predated the outer structure, having been built and then used by Edmond Halley and others since the 1720s. There were, in addition, observing rooms of less regular use placed to the south of the Quadrant Room and in the small matching 'summer houses' either side of Flamsteed House (Fig. 1).

A succession of assistants took up residence within these walls during Maskelyne's long tenure. Many of them stayed only a few months, some a couple of years and only one as long as seven. This high turnover seems to have been more to do with the demanding and yet dull nature of the work and life at the Observatory than because Maskelyne was difficult to work with. Although it seems that he would have liked to retain his assistants longer – and he hoped that raising salaries would achieve this – in many ways he made a virtue out of the pattern. Young men could receive an excellent training at Greenwich, which would stand them in good stead for future employment. As we will see, Maskelyne stayed in touch with most of them and later used some as computers, either to carry out the mathematical work required by the Observatory or to publish the *Nautical Almanac*, or as expeditionary observers.

Assistants had to be on call for observing duty between 7 a.m. and 10 p.m. every day of the week. This was not constant work but made it impossible to leave the Observatory. There was often night observing in addition, making for interrupted sleep, and a regular schedule of computation work to carry out in the evening when not observing. Maskelyne himself kept to a tough observing schedule except when absent in London for meetings, or either visiting his family or making his annual trip to the parish in which he was still a curate. He was not unaware of the demands of the job and in 1787 outlined the requirements for an assistant in a notebook:

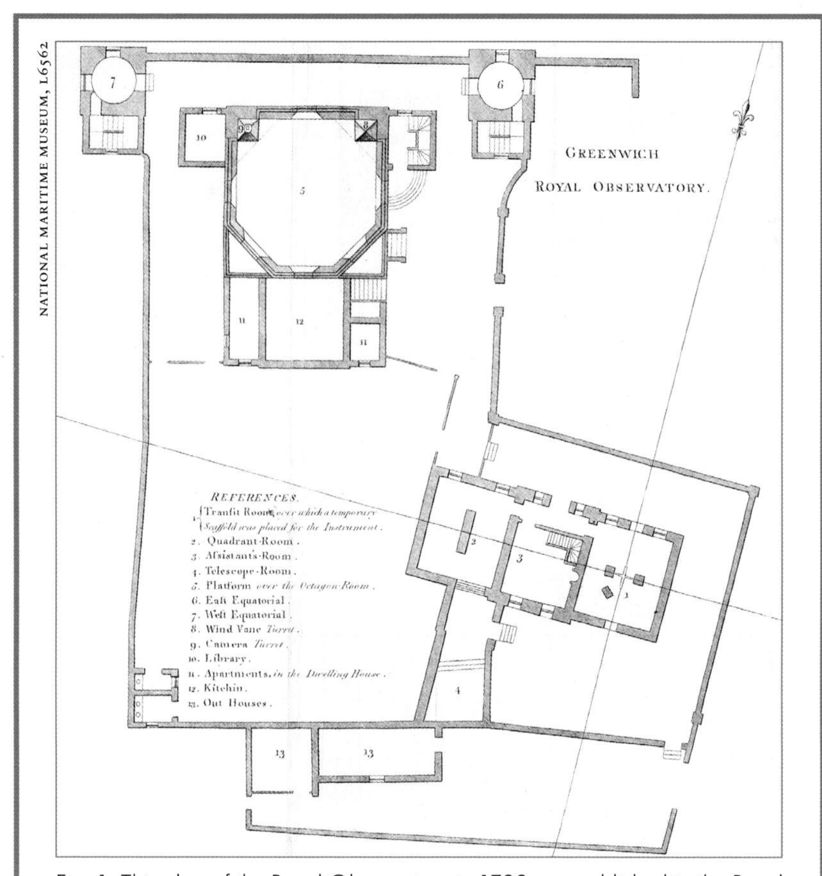

Fig. 1: This plan of the Royal Observatory in 1788 was published in the Royal Society's journal, *Philosophical Transactions*, in 1790, within General William Roy's report on the triangulation to establish the exact distance between the observatories of Greenwich and Paris. The residence of the Astronomer Royal and his family, Flamsteed House, is at the top, with the summer houses on either side. The meridian observatory and rooms of Maskelyne's assistant are shown to the right.

To understand Arithmetic, Geometry, Algebra, Plane & spherical trigonometry, & Logarithms; to have a good eye & good ears, be well grown, & have a good constitution to enable him to apply several hours in the day to calculation, & to get up to the observations that happen at late hours in the night. To write a good hand, and be a ready & steady arithmetical computer. If he know something of Astronomy

> & had a mechanical turn so much the better. To be sober & diligent, & able to bear confinement. Age from 20 to 40.[2]

If the arrangement did not seem to be working, Maskelyne was happy to end it, although he continued to feel a sense of responsibility towards former employees. The same is true of his relationship with *Nautical Almanac* computers, as is shown in the following chapter, where we meet David Kinnebrook, who was a one-time assistant who later took on some computing. Kinnebrook's letters to his father are a window onto life at the Observatory and show Maskelyne as someone who, though always concerned for its efficient running, was also anxious to help where he could and become a friend if possible.

When an assistant had been in position for a while, Maskelyne was able to trust him to continue work for periods by himself, although this led to a lonely life. Thomas Evans, the assistant who followed Kinnebrook in 1796, wrote:

> Nothing can exceed the tediousness and ennui of the life the assistant leads in this place, excluded from all society, except, perhaps, that of a poor mouse which may occasionally sally forth from a hole in the wall, to seek after crumbs of bread dropt by his lonely companion at his last meal. This, of course, must tend very much to impede his acquiring astronomical information, and damp his ardour for those researches which conversation with scientific men never fails to inspire. Here forlorn he spends days, weeks, and months, in the same long wearisome computations, without a friend to shorten the tedious hours, or a soul with whom he can converse.[3]

The assistants were able to leave the Observatory on Sundays and occasionally other days. They were also not completely alone up on Greenwich Hill, even if Maskelyne was in town. From the 1780s Maskelyne's wife and daughter were also based in Flamsteed House, along with a nursemaid or governess and four servants: 'a Lady's Maid, House Maid, Cook, and Footman'.[4] The assistants were in some ways part of the household. Kinnebrook reported that the footman dressed his hair two or three times a week and cleaned his shoes, that he was allowed to read books from Maskelyne's library and the London papers that he took, and that of the previous assistants 'some had dined with [Maskelyne] and some not'.[5] Whether Evans chose to do so or not, he was able to become close to someone in Maskelyne's household, since he later married the governess.

There were also visitors to the Observatory. Kinnebrook reported that he had 'seen several Gentlemen eminent in the Mathematical line since I have been here; amongst whom are the following Rev^{d}. M^{r}. Brinkely Rev^{d}. M^{r}. Woollaston Rev^{d}. D^{r}. Shepherd and M^{r}. Wales', adding that 'M^{r} and M^{rs}. Vine were also here ... and staid two or three Days'.[6] Other visitors included William Herschel, his wife, Mary, and sister Caroline. For both William and Caroline, Maskelyne had acted to confirm discoveries of comets or, most significantly, of a new planet, Uranus. They were also welcome dinner guests, William's reported remark on first meeting Maskelyne – 'That is a devil of a fellow' – evidently being one of approbation.[7] The Herschels and the Maskelynes, together with the Kellys and others, seem to have been regular companions. Caroline Herschel evidently became very fond of Maskelyne, who referred to her as 'my worthy sister in astronomy'.[8]

Even when over sixty, according to Kinnebrook, Maskelyne appeared to be 'a very healthy Man, and remarkably active', focused on his work and correspondence, and with regular business to attend to in London.[9] He was in regular contact with his wider family, discussing personal and financial matters and enjoying produce sent from their estates. We know relatively little about his personal philosophies or religious observance, which does not seem to have been demonstrative in nature. Kinnebrook noted that 'D^{r}. M. does not preach any where when he is at Home but he has a Living in Wiltshire where he generally preaches during the time of his stay in Wiltshire'.[10] It is notable that, on his death, his daughter chose not to emphasize her father's religious calling or belief in revealed religion on his memorial tablet, as her father's friend Samuel Vince urged her: it instead simply recorded that he 'worshipped the Great Creator/ By formulating laws of Nature', and that, 'Virtuous without pretence/He demonstrated goodness in his work'.[11]

4

NEVIL MASKELYNE AND HIS HUMAN COMPUTERS

Mary Croarken

INTRODUCTION

The creation and annual publication of the *Nautical Almanac* was one of Maskelyne's significant achievements during his time as Astronomer Royal. Harrison's watch was a potential solution to the longitude problem but until the early decades of the nineteenth century such technology was unaffordable to most sailors, beyond some captains of the East India Company or other successful trading ventures. The *Nautical Almanac* supplied an alternative method of finding longitude using astronomical observation taken on board ship, astronomical data supplied in table form and some mathematical calculations to combine the two. By publishing the *Nautical Almanac* sufficiently far in advance, and by selling it in ports around the trading world, Maskelyne provided a reliable and cheap method (2s 6d per annum) by which seamen could find their longitude astronomically – thus supporting both the exploration, colonial and trading needs of Britain during the second half of the eighteenth century and beyond.[1]

This chapter will describe how Maskelyne developed the ideas and techniques used in the *Nautical Almanac*, how he organized a workforce of 'human computers' to prepare the tables that it comprised and, briefly, what happened to it after his death.

To publish the *Nautical Almanac* on an annual basis took a substantial amount of organization and attention to detail, and a significant workforce to create the tables of astronomical data required. Between his appointment as Astronomer Royal in 1765 and his death in 1811, Maskelyne employed thirty-five individuals to undertake the calculation and preparation of the tables. The stories of these people, Maskelyne's human computers, and their interaction with him, demonstrate a different side to Maskelyne's character than that commonly portrayed as part of the Harrison story.

USING THE MOTION OF THE MOON TO FIND LONGITUDE

In 1755 Tobias Mayer, an astronomer from Göttingen, produced tables that gave the predicted position of the Moon much more accurately than any previously available. Mayer had based his lunar tables on recent theoretical work by the Swiss mathematician, Leonhard Euler. As discussed in Case study A, these more accurate tables presented an opportunity to calculate longitude using the lunar-distance method.

During his voyage to St Helena to observe the 1761 transit of Venus, Maskelyne began seriously to develop ideas on the practicality of finding longitude using lunar distances and, at the same time, to test how good Mayer's tables really were. The results were excellent and after returning home he published *The British Mariner's Guide* (1763), which gave instructions on how to use the technique and some of the required tables. He used real worked examples to clearly demonstrate his techniques in much the same way that his assistant on the voyage, Robert Waddington, did in his *Epitome of Theoretical and Practical Navigation*. The main problem with the technique was that the calculations took almost four hours to complete, something not likely to be popular with the average navigator on board a merchant or British naval ship.

Meanwhile, Harrison's watch (H4) was being trialled with equally promising results and, in 1764, there was a second trial of it on a voyage to and from Barbados. The Board of Longitude representative on the latter occasion was Maskelyne, who also used the journey to test an improved set of lunar tables produced by Mayer. On his return Maskelyne was appointed Astronomer Royal and reported to the Board of Longitude on both Harrison's watch and the lunar-distance method of calculating longitude at sea. The Board decided to award Harrison £10,000, with the remainder of the £20,000 longitude reward to depend on disclosure of the workings of the watch and the production of replicas to the same standard. As a result of the success of Maskelyne's practical application of Mayer's tables to finding longitude using 'lunars', the Board also awarded Mayer's widow £3,000 in recognition of her husband's work. Euler was awarded £300 in recognition of the mathematical theorems he had developed, which had made the production of Mayer's lunar tables possible.

At the same Board meeting, Maskelyne proposed that, in order to make the lunar method more practical to the everyday seaman, the predicted lunar-distance tables given at three-hourly intervals should be published annually along with all the other astronomical tables needed by navigators, such as the positions of stars, solar tables, predicted eclipses of the moons of Jupiter, positions of the planets and so on. This would reduce the calculations needed on board ship from four hours to approximately thirty minutes for each longitude calculation – a significant improvement. The Board agreed, committing itself to funding and publishing such tables for a trial period, and tasked both Maskelyne and the Oxford and Cambridge academics among its number 'to "look out for proper persons to calculate different parts" of what became the *Nautical Almanac*, and to report to the next Board the terms on which such persons will respectively undertake to make the said calculations'.[2]

Two weeks later four such persons were hired to compute the *Nautical Almanac* for 1767 and 1768. They were Israel Lyons (a well-known Jewish mathematician, astronomer and botanist whose religion precluded him from study at Oxford or Cambridge), George Witchell (soon to be headmaster at the Royal Naval Academy, Portsmouth), Williams Wales (an astronomer and teacher) and John Mapson (a less well-known mathematical assistant acquainted with Charles Mason, the current assistant at the Royal Observatory). These four soon got to work to calculate and produce the required collection of tables and Maskelyne published the first edition of the *Nautical Almanac and Astronomical Ephemeris* in 1767. The second edition for the year 1768 followed very closely. The *Nautical Almanac* was well received and the Board of Longitude agreed to continue the work. Although other almanacs were available the *Nautical Almanac* was the first to publish lunar distances for the express purpose of calculating longitude and it soon gained an international reputation for high accuracy and reliability. It has been published annually ever since, although today it is in a much different format. Maskelyne received no financial reward for the work: its ongoing production was very much seen as an integral part of his role as Astronomer Royal (Fig. 1).

THE

NAUTICAL ALMANAC

AND

ASTRONOMICAL EPHEMERIS,

FOR THE YEAR 1767.

Publiſhed by ORDER of the

COMMISSIONERS OF LONGITUDE.

LONDON:

Printed by W. RICHARDSON and S. CLARK,
PRINTERS;
AND SOLD BY
J. NOURSE, in the Strand, and Meſſ. MOUNT and PAGE
on Tower-Hill,
Bookſellers to the ſaid COMMISSIONERS.
M DCC LXVI.

Fig. 1: The title page of the first edition of the *Nautical Almanac and Astronomical Ephemeris*, providing tables for the year 1767.

CALCULATING THE *NAUTICAL ALMANAC*

It was no trivial task to calculate all the tables in the *Nautical Almanac* for a whole year at a time. It needed a reasonable workforce but, just as importantly, required all those involved to be working in the same way and to the same level of accuracy.

The first *Nautical Almanac* set a style that endured until 1834 and was arranged into four sections:

- the preface, which explained the purpose of the publication and the resources used to calculate the tables;
- a key to the symbols used within the tables and a list of expected eclipses and other astronomical data;
- the monthly tables themselves;
- an explanation of how to use the tables, often with examples.

The tables were arranged by calendar month with twelve pages per month. Included were tables of the Sun's longitude, the timings of the eclipses of Jupiter (which could be more accurately used to determine longitude on land), the positions of the planets, the position of the Moon for noon and midnight for every day of the month, and the lunar distances from the Sun and stars for every three hours of the day (Fig. 2).

Any one *Nautical Almanac* month required 1,365 entries to be calculated. To undertake this work on an ongoing basis, Maskelyne employed a series of 'computers' – i.e. people who 'computed' or carried out the necessary calculations. While the initial computers – Lyons, Mapson, Wales and Witchell – did a good job in testing out the feasibility of Maskelyne's plans, they all had other roles and could not commit to this relatively menial task full-time. As already mentioned, between 1767 and 1811 there were thirty-five individuals who acted as *Nautical Almanac* computers for Maskelyne. Regardless of how long it took them or what corrections they needed to make, the computers were paid per *Nautical Almanac* month computed (starting at £5 16s 8d in 1767 and rising with the inflation caused by the Napoleonic

APRIL 1767.

Diſtances of ☽'s Center from ☉, and from Stars weſt of her.

Days.	Names.	12 Hours. ° ′ ″	15 Hours. ° ′ ″	18 Hours. ° ′ ″	21 Hours. ° ′ ″
1	The Sun.	40. 59. 11	42. 34. 44	44. 9. 51	45. 44. 35
2		53. 32. 7	55. 4. 24	56. 36. 16	58. 7. 45
3		65. 39. 18	67. 8. 27	68. 37. 14	70. 5. 39
4		77. 22. 36	78. 48. 58	80. 15. 1	81. 40. 46
5		88. 45. 20	90. 9. 27	91. 33. 21	92. 57. 0
6		99. 52. 6	101. 14. 34	102. 36. 52	103. 59. 1
7		110. 47. 42	112. 9. 6	113. 30. 25	114. 51. 40
6	Aldebaran	50. 36. 10	52. 4. 5	53. 31. 57	54. 59. 44
7		62. 17. 43	63. 45. 10	65. 12. 34	66. 39. 57
8	Pollux.	31. 25. 48	32. 53. 11	34. 20. 40	35. 48. 12
9		43. 7. 5	44. 35. 4	46. 3. 8	47. 31. 15
10	Regulus.	17. 51 57	19. 20. 36	20. 49. 26	22. 18. 27
11		29. 45. 36	31. 15. 26	32. 45. 26	34. 15. 35
12		41. 48. 49	43. 19. 55	44. 51. 10	46. 22. 36
13		54. 2. 11	55. 34. 36	57. 7. 12	58. 39. 59
14		66. 26. 28	68. 0. 18	69. 34. 20	71. 8. 33
15	Spica ♍	25. 4. 34	26. 39. 23	28. 14. 26	29. 49. 44
16		37. 49. 37	39. 26. 14	41. 3. 5	42. 40. 8
17		50. 48. 40	52. 26. 59	54. 5. 31	55. 44. 15
18		64. 1. 2	65. 41. 3	67. 21. 18	69. 1. 48
19	Antares.	31. 37. 14	33. 19. 7	35. 1. 13	36. 43. 32
20		45. 18. 29	47. 2. 10	48. 46. 5	50. 30. 12
21		59. 14. 6	60. 59. 31	62. 45. 11	64. 31. 2
22		73. 23. 37	75. 10. 43	76. 58. 2	78. 45. 31
23	β Capricorni.	33. 17. 26	35. 4. 38	36. 52. 4	38. 39. 45
24		47. 41. 9	49. 29. 53	51. 18. 44	53. 7. 40
25	α Aquilæ.	65. 57. 35	67. 29. 54	69. 2. 36	70. 35. 39
26		78. 24. 51	79. 59. 9	81. 33. 29	83. 7 45

Fig. 2: A page from the first edition of the *Nautical Almanac*, showing predicted lunar distances for April 1767. These could be used to establish the time in Greenwich and, with the correct observations and calculations, to establish longitude.

wars to £18 15s 0d by 1811). Individual computers took from three to twelve weeks to do the work allocated to them depending on their work rate and other commitments. Very few worked full-time on the *Nautical Almanac* but, rather, used it to supplement other forms of income. Today we would recognize the workforce as a network of freelance homeworkers.

As there were so many individuals used to calculate different parts of the publication, it was important that they all worked in the same way: otherwise there was a risk of considerable variation in the tables from month to month and year to year. There was a significant amount of astronomical theory behind some of the tables, but Maskelyne did not expect each computer to be highly knowledgeable in computational astronomy. He needed people who were familiar with astronomical terms, good at using logarithmic tables to undertake calculations and not intimidated by having to undertake arithmetic in sexagesimal (i.e., degrees, seconds and minutes) rather than decimal notation.

To ensure that everyone was working to the same standards, Maskelyne developed a series of algorithms (or precepts) – basically a step-by-step set of instructions for each entry that was required. Each calculation could be broken down into a series of 'table look-ups' (i.e., finding an entry in another book of either astronomical or logarithmic tables) and a series of additions and subtractions. The use of logarithms meant that multiplication and division, the most difficult of the arithmetical operators, were reduced to more easily worked addition and subtraction. For example a calculation for any one entry for a lunar distance in the *Nautical Almanac* may require up to twelve table look-ups and fourteen seven- or eight-figure sexagesimal operations.

When new computers started working for him, Maskelyne would write to them describing the process and send a set of twelve to fourteen books of tables to use in undertaking the work.[3] The set of books included tables of lunar, solar and planetary motion, logarithmic tables, tables of products and powers of numbers, past copies of the *Nautical Almanac* and the accompanying publication *Tables Requisite to be used with the Nautical*

Almanac, and a set of unpublished tables and instructions produced by Maskelyne and his assistant at the Royal Observatory.[4] If a computer had any questions then Maskelyne would write to clarify any such issues (Fig. 3).

Greenwich Sept. 30. 1788

Sir,

In answer to your letter of the 23.d instant.
Compute D's dist. from a star by Logarithms, thus
P, the pole of the ecliptic, N or S. PS, PL are
the star's & moon's distances from one & the same pole.
L the D, S the star, SD a perpendicular
great circle let fall from S to PL, perpendicular
to it in D. There suppose P to be less than 90°.
By Log.s c, P + t, PS = t, PD & PL − PD = LD.
c, LS = − c, PD + c, PS + c, LD
or = c, P + s, PS − s, PD + c, LD
The latter formula must be used when PD
is large, or near 90°; but may be used safely
in all cases. —

Example

~~P 20.30 — c — 9.9715876~~
~~PS 89.58.30 t — 13.3857588~~
P 25.25.30 .. c -- 9.9557589 .... 9.9557589
PS ...... t 12.0977246 ...... s.. 9.9999879
PD 89.29.36 t 12.0534829 ---- s. co-ar. ---- 0.0000170
PL 85.0.0 ...... c. ---- 9.9986631
LD 4.29.36 ...... c --- 9.9544269
SL 25.47.30

If P should be greater than 90°
then PL + PD = LD as in this scheme

As the planets places, excepting the
sun and moon are only set down to minutes,
there is no occasion to allow for nutation &
aberration. — In computing the occultations of
stars by the D you must calculate their Longitudes & Lat.s
to seconds, and apply nutation & aberration. I am glad to
hear of the improvement of the town of Cambridge. I am
Your humble Servant
N. Maskelyne

P.S. Mr. Taylor has printed
to the 33° of the Log. sines & tangents—

Fig. 3: A letter from Nevil Maskelyne to one of his *Nautical Almanac* computers, Joshua Moore, dated 30 September 1788. See below for more information about Moore's career.

It was essential that the work of the computers was correct, as an error in one figure in the tables could, if in the wrong place, result in an error of sixty miles or so in a longitude calculation – possibly the difference between life and death for a ship's crew. To try to ensure that the calculations contained as few as errors as possible, Maskelyne devised checking mechanisms as part of the process.

For most table entries he allocated two independent computers to do the calculation; to distinguish them, one was called the computer and the other was called the anti-computer. Maskelyne kept a record of who had been assigned to what month (both so he knew to whom the work was assigned but also so he could pay for work completed), when he sent work to them and when the completed work was received (Fig. 4).[5]

The work itself was done several years in advance to ensure that the *Nautical Almanac* could be published in good time to be taken on lengthy voyages: ships could carry editions for several years ahead and Cook is known to have taken printed pages of as yet incomplete editions on his voyages of exploration.

Once the work for any one month had been allocated, computed and then returned to him, Maskelyne asked a third person, the 'comparer', literally to *compare* the two manuscripts and look out for any discrepancies. The comparer also did some spot checks and used another mathematical technique

1803	Computers of the Nautical D. Place at Noon	Almanac 1803 D. Place at Midnight
January	Mrs Edwards Dec. 8. 1791 Rec. Jan 31. 1792 [illegible]	Mr Moore Nov. 23. 1791 Rec. Feb. 4. 1792
February	Mrs Edwards Dec. 8. 1791 Rec. Jan 31. 1792 [illegible]	Mr Moore Nov. 23. 1791 Rec. Febr. 18. 1792
March	Mrs Edwards Jan. 31. 1792 Rec. Mar 20 (93)	Mr Moore Feb. 4. 1792 Rec. March 17 (92)
April	Mrs Edwards Jan. 31. 1792 Rec. Mar. 20 (93)	Mr Moore March 30. 1792 Rec. Apr. 19
May	Mrs Edwards March 20 (92) Rec. April 17	Mr Andrews March 10 (92) Rec. May 18
June	Mrs Edwards March 20 (92) Rec. April 17	Mr Andrews March 10 (92) Rec. July 9
July	Mr Moore March 30 (92) Rec. May 8	Mrs Edwards Apr. 17 (92) Rec. May 28
August	Mr Moore Apr. 30 (92) Rec. May 22	Mrs Edwards Apr. 17 (92) Rec. May 25
September	Mr Moore May 23 (92) Rec. Sept. 19	Mrs Edwards May 29 (92) Rec. July 13
October	Mr Moore May 23 (92) Rec. Oct. 10	Mrs Edwards May 29 (92) Rec. July 13
November	Mr Andrews July 10 (92) Rec. Sept. 18	Mrs Edwards July 14 (92) Rec. Sept. 2
December	Mr Andrews July 10 (92) Rec. Oct. 12	Mrs Edwards July 14 (92) Rec. Sept. 2
1804 January	Computers of the Nautical Almanac Mrs Edwards Sept. 3 (92) Rec. Oct. 27	of 1804 — N. 2500 ordered to be printed Mr Moore Oct. 23 (92) Rec. Dec. 7

Fig. 4: Maskelyne's Nautical Almanac Record Book, showing the allocation of work done in late 1793, for the calculation of the 1804 *Nautical Almanac*.

called differencing to ensure the correctness of the calculations. The computers would take the calculations to a certain point, then send them directly to the comparer, who checked them and returned them before the final stage of the computations were undertaken by the computers, which would in turn be returned to Maskelyne before he passed them to the comparer again. This ensured that any errors made in the early stages of the calculations did not survive through to the final tables. The comparer also had the task of proofreading the tables when they came back from the printers.

Although, today, double calculation is no longer considered a foolproof way of checking computations, since the likelihood of the same error being made twice is not insignificant, Maskelyne managed the system very successfully and the *Nautical Almanac* had a high reputation for correctness.

As an illustration of the thoroughness of the system and Maskelyne's determination to ensure the highest standards, take the example of Joseph Keech and Reuben Robbins. In 1770 the comparer, Malachy Hitchins, discovered that these two computers had copied each other's work when they had been assigned to the same *Nautical Almanac* month: their calculations were too much alike for them to have been independently computed. Keech and Robbins were part of the same London coffee-house set and lived relatively near to each other: it must have seemed like an easy fraud to have shared the work and made two copies. When their duplicity was discovered, not only did Maskelyne not pay the computers for the work in question but he also deducted a charge for the comparer's time from money he owed them for other work.

The comparer, Malachy Hitchins, continued to detect errors and support computers to improve their work. We have evidence of this process in the form of a series of letters, preserved in the Library of Congress in Washington D.C., between Hitchins and Joshua Moore, then a young computer, in which Hitchins points out a series of corrections (Fig. 5).

The Robbins-Keech experience taught Maskelyne never to assign work to two geographically close computers and, indeed,

Sir — Marazion, 19 Decr. 1788 —

The probable Phenomena in Decr 1798
are these, Days
4 — α ♎
19 — ☽ ♃
26 — ♄ ♌
31 — α ♎ Compute tho' in the daytime * being 2d Magn.

Be pleased to correct Equn. of time 1st Day
viz 10.16,9 — 8 23 to 31 ☉'s Long. the
15th Day I take to be an Error in transcribing.
Cor. Em. 1st Sat 29 12 59 13
All 2d Sat. wants Correction 1.13.32.35 &c.
Send ☿'s Gr. Elongn. in Decr. & Correct the
Southing of ♅ 18.15 &c.
☽'s Long. 1st Noon read 5 18 58.5
21st Noon read 2 9 35 25

You will have materials for Jany
1799 in a few Days from,

Sir, your very humble Servt
Malachy Hitchins

Fig. 5: A letter dated 19 December 1788 from the *Nautical Almanac*'s comparer, Malachy Hitchins, to computer Joshua Moore, pointing out a series of corrections.

from that time onward he actively sought out people who did not live in close proximity and were not acquainted with each other. He employed computers across the length and breadth of England, from Ludlow in the west to Norwich in the east and from Northumberland down to Cornwall. They all worked in their own homes and only came to London to collect their wages.

Some did not even do that, but arranged instead for Maskelyne to pay their London bills directly or had trusted relatives or friends collect the money on their behalf.

All communications between Maskelyne and his computers was by post, which was surprisingly efficient in the late eighteenth century. Some people worked for him for just a couple of months and some for nearly forty years.

THE COMPUTERS AND COMPARERS OF THE *NAUTICAL ALMANAC*

The individuals who made up the computers and comparers of the *Nautical Almanac* came from a variety of backgrounds, had different motivations for their employment by the Board of Longitude and different relationships with Maskelyne. Some were established astronomers with whom he was acquainted through their work for the Board – or indeed as his assistants at the Royal Observatory. Others were clergymen, schoolteachers or surveyors. In one case a whole family became involved. To demonstrate the diversity of the eighteenth- and early nineteenth-century *Nautical Almanac* computers, a few of their stories are told below.

ASTRONOMER: WILLIAM WALES (1734–1798)

Wales is perhaps the most well-known of Maskelyne's *Nautical Almanac* computers and one of the first he hired in 1765 to undertake the initial computations (Fig. 6).

We do not have a clear picture of how Wales became known to Maskelyne, but he had come to London from his home in Wakefield, Yorkshire, in the 1750s with an interest in mathematics and astronomy. He was a contributor to the well known *Ladies' Diary*, an annual almanac that included mathematical puzzles with reader solutions. In 1765 Wales married Mary Green, the sister of Charles Green, astronomical assistant at the Royal Observatory. Wales was therefore part of the small astronomical community in London in the 1760s and may have been recommended for the work, which seems to have kick-started his career in computational and observational astronomy.

Although one of the first computers, Wales did not remain as a permanent one. In 1768 he was sent by the Royal Society to Hudson's Bay to observe the 1769 transit of Venus, his instructions being given to him by Maskelyne. On his return Maskelyne provided him with more *Nautical Almanac* work before he sailed as astronomer on Cook's second voyage to the South Pacific in 1772–5.

On his return Wales was appointed Master at the Royal Mathematical School at Christ's Hospital but thereafter, when he needed a senior person, Maskelyne occasionally asked him to help out as a comparer. Maskelyne and Wales were both Fellows of the Royal Society and moved in the same circles in the latter part of the eighteenth century. Wales was therefore a safe pair of hands as far as Maskelyne was concerned. He relied on him to start the computing work and then, when he required expertise in later years, knew he could be called upon. For Wales, computing the *Nautical Almanac* was first a stepping-stone to working as a Board of Longitude astronomer and later a source of extra income.

Fig. 6: A pastel portrait of William Wales by John Russell, 1794. By this date Wales was Master of the Royal Mathematical School at Christ's Hospital.

Wales's brother, John, who did not come to London looking for adventure but remained in Wakefield, also worked for Maskelyne on an *ad hoc* basis, contributing to eighteen editions of the *Nautical Almanac*. William sometimes collected payments from Maskelyne on his brother's behalf. Maskelyne's relationship with William and John Wales is an indication of how small a circle astronomers moved in and how recommendation through familial channels was a standard way of finding skilled people – or, perhaps, people who could be trusted and then taught.

ASTRONOMER: JOHN CROSLEY (1762–1817)

Another Board of Longitude astronomer and sometime *Nautical Almanac* computer was John Crosley who, like Wales, came from Yorkshire but was somewhat younger (Fig. 7).

From 1789 to 1792 Crosley worked as Maskelyne's assistant at the Royal Observatory, Greenwich. This role involved making regular observations, often at night, reducing these observations (i.e. taking the mean of a series of short-interval observations and making the necessary corrections to them) and undertaking a variety of astronomical computations under Maskelyne's guidance. In 1793 Crosley was employed by the Board of Longitude as an astronomer accompanying George Vancouver's exploration of British Columbia, in the *Discovery*. This was the first of several Board-supported voyages for Crosley, all of which involved discovery, survey expeditions, and chronometer testing. During his explorer career he experienced shipwreck and being left behind at the Cape of Good Hope after falling ill in 1802.

Following his adventures at sea, Crosley settled in London and earned a living doing various bits and pieces of computing for the Board (including *Nautical Almanac* calculations around 1805 and 1810). It is clear that computing was an important part of Crosley's income from 1799 onwards, both between voyages and after his travels came to an end in the early 1800s. Crosley served as President of the Spitalfields Mathematical Society from 1805 until his death in 1817.

Fig. 7: An engraved portrait of John Crosley, included in a scrapbook that was presented to the Spitalfields Mathematical Society in 1813, when Crosley was its president.

ASTRONOMICAL ASSISTANT: DAVID KINNEBROOK (1772–1802)

Maskelyne employed several ex-Royal Observatory assistants on an occasional basis as *Nautical Almanac* computers: one, however, stands out. David Kinnebrook was born in 1772, in Norwich, where his father was a schoolmaster. In May 1794, Kinnebrook came to Greenwich as Maskelyne's assistant but was dismissed less than two years later because his recorded observations were consistently eight-tenths of a second different from Maskelyne's own.

In the eighteenth century, observations were taken by noting the time on the observatory clock and then counting the seconds by listening to the tick of the clock while watching an astronomical body passing through the lines in the telescope viewfinder. The identification of the required tenth-of-a-second timing therefore necessitated both judgement and skill. In 1820 the German astronomer Freidrich Bessel identified that different individuals had natural differences in perception, which should and could be accounted for by a 'personal equation' when making astronomical observations under observatory conditions. However, in the 1790s, Maskelyne was unaware of this phenomenon.

Kinnebrook was devastated to be 'let go' from the Royal Observatory, although his letters home to his father indicate that he was far from happy at Greenwich. He was lonely and isolated living and working in the Observatory and having to be on hand almost '24/7', but the increasing pressure over the 'incorrectness' of his observations was also telling.[6] Kinnebrook found the next few years hard. He obtained work as a not-very-successful schoolmaster at Gresham School in north Norfolk where he lived in great fear of being conscripted to fight in the French Revolutionary War. By 1801, when his health was failing and he could no longer stand the rigours of boisterous schoolboys, he wrote to Maskelyne asking for work. Maskelyne, revealing the compassionate side of his character, supplied him with two months' worth of *Nautical Almanac* computing work just before Kinnebrook's death in April 1802, aged only thirty.

ASTRONOMER AND CLERGYMAN: REVEREND MALACHY HITCHINS (1741–1809)

The Reverend Malachy Hitchins was a key part of the success of the *Nautical Almanac* computations for many years. His story illustrates both the way in which Maskelyne used his existing contacts and the fact that he wanted to ensure that those working on the *Nautical Almanac* lived some distance from each other.

Hitchins was from Cornwall and, when a youth, worked as a miner. He started his mathematical career working as a surveyor with Benjamin Donne on the latter's 1765 map of Devonshire, before going to Oxford University in 1763 and taking holy orders in 1767. Becoming a clergyman was important to Hitchins as a way of providing a steady income with which to support his family but, in parallel to this activity, he acted as a temporary assistant at the Royal Observatory in 1768–69 while the full-time assistant, William Bayly, travelled to Norway to observe the 1769 transit of Venus. Hitchens contributed to the transit observations in England.

Maskelyne soon came to value Hitchins' qualities and got him involved in the *Nautical Almanac*, first as a computer in 1767 and, soon afterwards, as the comparer – a role he played until his death in 1809. He acted as Maskelyne's right-hand man for forty years – directing, teaching and supervising the computers as well as double-checking all their work. He did all of this from the small village of St Hilary in Cornwall, where he became vicar in 1775. As already noted, it was Hitchins who uncovered the cheating by Keech and Robbins.

Once again Maskelyne used an ex-Observatory assistant with whose qualities he was familiar. Hitchins, in turn, sometimes recommended other computers to Maskelyne. For example, Hitchins was connected to a famous family of surveyors in Cornwall, the Martyns and Pascos, and recommended various members of the family across Devon and Cornwall. Nicholas James was the schoolmaster in St Hilary, where Hitchins lived, and was brought in by him to cover when other computers had let Maskelyne down (or, in a couple of cases, proven to be inept). By 1801, however, James was employed as computer in his own right, with the subsequent increase in income supporting his growing family. William Dunkin was another local St Hilary man who began computing for Maskelyne in 1808, just before Hitchins died, but who continued in the post until 1838. Dunkin's sons both had careers in computational astronomy; Edwin, as an assistant at the Royal Observatory, while Richard worked for the Nautical Almanac Office.

CLERGYMAN AND SCHOOLTEACHER: THOMAS BROWN (1755–1836)

Another clergyman involved with the *Nautical Almanac* was Thomas Brown. Brown was a Derbyshire clergyman, also headmaster of Tideswell Grammar School, and was related by marriage to William Lax, Lowndean Professor at the University of Cambridge and a Commissioner of Longitude. In January 1811, just before he died, Maskelyne was persuaded to take on Brown as comparer, as a permanent replacement for Hitchins. It did not turn out well. Brown had neither the meticulousness nor dedication to make a good job of the task and he was hugely committed as headmaster at Tideswell.[7] He also lacked the experience to know what was needed both in computational and management terms. When Maskelyne died leaving the *Nautical Almanac* in his hands (as John Pond, the subsequent Astronomer Royal, had little interest in the work), the quality declined rapidly to the point of complaints being raised in Parliament some years later.

CLERGYMAN AND INSTRUMENT MAKER: JOHN EDWARDS (C.1747–1784)

Edwards was a more successful clerical appointment. Following undergraduate study at Cambridge, he was ordained into the Church of England and obtained several livings in his home county of Shropshire. Edwards' real interest was not theology but astronomy – in particular making and improving telescope mirrors, in which field he was a leading expert and active researcher. Acquiring the materials to undertake his telescope-mirror research was as expensive as it was time-consuming. To supplement his clergyman's income and support his family, Edwards took in pupils preparing for Cambridge but this was not enough to cover his mounting research costs.

In 1773 Edwards was introduced to Maskelyne, probably by Edward Waring, Lucasian Professor of Mathematics at Cambridge and a Commissioner of Longitude, who knew Edwards through their Shropshire and Cambridge connections. Maskelyne offered Edwards work as a computer of the *Nautical Almanac* and this helped to pay for his telescope-mirror work (as did two awards from the Board of Longitude of £20 and £200 in 1778 and 1780 respectively), especially as bills from London instrument makers and raw-material suppliers could be directly settled by Maskelyne from his *Nautical Almanac* earnings. In addition, Maskelyne also published a paper written by Edwards on telescope mirrors as an annexe to the 1781 edition.

Edwards lived within twenty miles of Maskelyne's sister Margaret, Lady Clive, and it is obvious that Maskelyne knew the Edwards family very well. Maskelyne's memorandum book for 1780–81 includes cures for ague and tips for calming horses that he received from Edwards.[8] The relationship sounds close and warm.

In 1784 John Edwards died from fatally inhaling the arsenic he was using to help refine the metals for his telescope mirrors. He was obviously a well-known member of his local community as the *Shrewsbury Chronicle* published a notice of his death:

> Last Saturday was interred, the remains of the Rev. Mr. John Edwards of Ludlow – A gentleman of extensive learning and knowledge in Philosophy, Astronomy, and most Branches of Mathematics. He, for several years, (with great credit to himself) had assisted the Rev. Dr. Nevil Maskelyne, the Astronomer Royal, in calculating the Nautical Ephemeris for the Board of Longitude.[9]

The notice was accompanied by a florid poetical tribute with more astronomical references than the local population would normally have seen in their weekly news-sheet.

John Edwards left behind his wife Mary and young children Eliza, Maria and William. Not only was Mary husbandless and homeless (their house [Fig. 8] being tied to John's position in the church) but also without a regular income.

PRIVATE COLLECTION

Fig. 8: The preacher's house in Ludlow, where John and Mary Edwards lived before his death in 1784.

FULL-TIME COMPUTERS: MARY EDWARDS (C.1750–1817) AND ELIZA EDWARDS (1779–1846)

Following John's death, Maskelyne's accounts and the *Nautical Almanac* record books show that the name of John Edwards is replaced, with no comment or fuss, by Mary's name: yet it was rare in England that a woman would undertake this type of paid work for a public body such as the Board of Longitude. There are examples of women doing computation work as an anonymous helpmate to a husband or brother, but to be paid by the state was unusual. From later letters from Mary to the Board of Longitude, following Maskelyne's death, we know that – all along – Mary had been carrying out *Nautical Almanac* computations while her husband concentrated on his telescope-mirror research[10]. As demonstrated above, Maskelyne knew the family well and was undoubtedly already aware that Mary had been doing most, if not all, of the computing for which John was being paid. As long as the work was done to a high standard Maskelyne was content.

After her husband's death, Mary leased a new home in Brand Street, Ludlow, and for forty years earned her living as a computer for the *Nautical Almanac*: she was the most prolific computer working full-time, doing almost half of the work for each *Nautical Almanac* edition. By 1810 Mary had saved enough to buy the house and secure a future for her daughters.

We do not know how Mary came to have the skills to be a computer. She was perhaps taught by her father, Thomas Harrison, an educated blacksmith from the village of Longnor just south of Shrewsbury, or by John himself. But in the same way that other women in Ludlow made gloves at home to keep their families, Mary Edwards was a computer working with pen, paper and books of log tables.

Mary's daughter, Eliza, began to help her mother when she was a young girl and continued to do the computations following her mother's death in September 1815. Eliza continued to work for the *Nautical Almanac* until 1832 when the work was centralized in London. She died in 1846 having never married, and bequeathed her mother's house to her faithful servant, Selina Jones.

SCHOOLTEACHER: HENRY ANDREWS (1744–1820)

Several of Maskelyne's computers were schoolteachers: William Wales, William Garrard, Charles Barton, a reluctant David Kinnebrook, Nicholas James and the infamous Reuben Robbins, to name a few. One of the most significant schoolteacher computers was Henry Andrews (Fig. 9).

Andrews was brought up in Lincolnshire, spent his early adulthood in Cambridge and moved to Royston in Hertfordshire in 1766, where he and his wife opened a mathematical school alongside a shop selling scientific instruments and books. He was probably introduced to Maskelyne by Richard Dunthorne, the first comparer of the *Nautical Almanac*, whom he had known while living in Cambridge. Andrews started computing for the *Nautical Almanac* in 1768 and continued until 1815, when he retired aged seventy-one.

Fig. 9: An engraved portrait of Henry Andrews, with a celestial globe. He is identified as an 'Astronomical Calculator to the Board of Longitude' as well as 'Author of Moore's Almanac'. This print by Thomas Blood post-dates Andrews' death in 1820, but the original on which it is based was by James Watson, who died in 1790.

For Andrews computing not only supplemented his income but also became a passion. From 1776 to 1819, alongside his *Nautical Almanac* work, he prepared the astronomical prediction tables for *Old Moore's Almanac,* an annual publication edited by Charles Hutton, Professor of Mathematics at the Royal Military Academy, Woolwich, and a friend of Maskelyne.

Letters between Maskelyne and Henry Andrews were generally formal and professional, not those of close friends.[11] Maskelyne's relationship with Andrews was not warm, as with the Edwards family, or built on a personal relationship like those with Wales, Crosley or Hitchins, although the correspondence discussed in the conclusion to Chapter 5 shows that their long association led to some familiarity between the two men and their families. It is clear, however, that Maskelyne valued Andrews as a steady computer and often settled accounts for him from his *Nautical Almanac* earnings for books and scientific instruments, at shops and with instrument makers across London.

BOARD OF LONGITUDE COMPUTER: MICHAEL TAYLOR (1756–1789)

Maskelyne often employed as computers people who worked in other capacities for the Board of Longitude. He was usually responsible for directing their work there and knew the skills that each of them possessed.

Michael Taylor, for example, made several significant tables for the Board. The most famous were his published *Sexagesimal Tables* (1780) and seven-figure *Logarithm Tables* (1792), but there were numerous unpublished tables as well that were used by the Royal Observatory, or for other aspects of Board work. In between commissions for table-making, Taylor was a *Nautical Almanac* computer. He was well known to Maskelyne and, after his premature death in 1789, Maskelyne used monies owed to him from the publication of his *Logarithm Tables* (£930) to ensure an education and future for Taylor's son. Again, Maskelyne is seen to be caring and compassionate in his dealings with individuals.

BOARD OF LONGITUDE COMPUTER: JOSHUA MOORE (ACTIVE 1787–1805)

Moore began his association with Maskelyne in 1787 by transcribing papers and observations for the Royal Observatory. He later moved on to *Nautical Almanac* computation while living at the Angel Inn at 17 Market Street, Cambridge. I have not managed to establish either why he had moved to central Cambridge (although he perhaps had a link to the Cambridge Observatory) or how else he made a living, but for just over five years he worked more or less full-time on *Nautical Almanac* calculations.

In the 1790s, computing for the *Nautical Almanac* was suspended for a period as preparation of the tables had got almost ten years ahead of their publication dates. During this time, Moore moved first to London before sailing to America on 8 October 1793. Surveyors were in demand in the Americas at this time but currently I have found no evidence of his life there other than records in the Library of Congress of him writing to Thomas Jefferson in 1805, advising on how to find longitude without a clock.

AUTHOR AND RADICAL: BENJAMIN WORKMAN (ACTIVE 1786–1818)

Benjamin Workman was a *Nautical Almanac* computer with a distinctly different background. He emigrated to America from Ireland before 1786, worked as an instructor in mathematics at the University of Pennsylvania and was the author of several mathematically based books. He was also active in the anti-federalism movement, publishing essays advocating the case and calling for a free local press. He was eventually dismissed from the University for his radicalism and then spent four years as a poorly paid ship's schoolmaster. He was discharged because of ill health – specifically an inflammation of the eyes – and eventually ended up in London.

Desperate for employment, Workman petitioned Maskelyne for work in 1809. Maskelyne did indeed allocate some computation to Workman but, whether his eyesight let him down or there was another reason, the work was not up to the standards Maskelyne expected and had to be repeated by another computer. Soon afterwards Workman found himself in Newgate prison for a debt of £16. Maskelyne obviously felt some sympathy for Workman as he helped to raise a small sum for food and drink for him in the prison.

SURVEYOR AND BOARD OF LONGITUDE COMPUTER: PHILLIP TURNER (1752–1800)

Another problematical computer for Maskelyne was Phillip Turner. In the late 1770s, Turner was employed as a surveyor by the Hudson Bay Company on the recommendation of William Wales but twenty years later he was back in England doing a variety of copying and computing work for the Board of Longitude. He began to undertake *Nautical Almanac* computing in May 1799 but died a couple of months later. Maskelyne had to work hard to retrieve the expensive books with which Turner had been supplied in order to complete the calculations and, in the end, did not get them all back. In his accounts Maskelyne speculates that Turner's landlord had sold them in lieu of unpaid rent.

THE COMPUTERS AND COMPARERS OF THE *NAUTICAL ALMANAC*, 1766–1811

NAME	COMPUTER/ COMPARER	RANGE OF EDITIONS WORKED ON*	GEOGRAPHICAL LOCATION WHILE A COMPUTER	OTHER OCCUPATIONS
Henry ANDREWS	Computer Comparer	1770–1815 1815	Royston, Hertfordshire	Schoolmaster Bookseller Tablemaker
Charles BARTON	Computer Comparer	1775–81 1770–86	Greenwich	Schoolmaster
William BAYLY	Computer	1778–89	Greenwich	Astronomer Schoolmaster Headmaster

NAME	COMPUTER/ COMPARER	RANGE OF EDITIONS WORKED ON*	GEOGRAPHICAL LOCATION WHILE A COMPUTER	OTHER OCCUPATIONS
Thomas BROWN	Comparer	1815–16	Tideswell, Derbyshire	Clergyman Headmaster
Richard COFFIN	Computer	1783–86	Devon (probably)	Uncertain
John CROSLEY	Computer	1805–16	Spitalfields, London	Astronomer Computer
William DUNKIN	Computer	1815–17	St Hilary, Cornwall	
Richard DUNTHORNE	Comparer	1767–76	Cambridge	Astronomer Surveyor Superintendent
Eliza EDWARDS	Computer	1810	Ludlow	
John EDWARDS	Computer	1773–84	Ludlow	Clergyman Instrument maker
Mary EDWARDS	Computer Comparer	1784–1810 1810	Ludlow	
George GILPIN	Computer	1779–95	London	Astronomical assistant Clerk
John HELLINS	Computer	1780	Devon	Clergyman
Malachy HITCHINS	Computer Comparer	1769–1814	St Hilary, Cornwall	Clergyman
Charles HUTTON	Comparer	1779–83	Woolwich	Professor of mathematics
Nicholas JAMES	Computer Comparer	1805–16	St Hilary, Cornwall	Schoolmaster
Joseph KEECH	Computer	1769–83	London	Clerk
David KINNEBROOK	Computer	1807–08	Norwich	Astronomical assistant Schoolmaster
Israel LYONS	Computer	1767–68	Cambridge	Botanist Astronomer
John MAPSON	Computer	1767–71	Tetbury, Gloucestershire	Uncertain

NAME	COMPUTER/ COMPARER	RANGE OF EDITIONS WORKED ON*	GEOGRAPHICAL LOCATION WHILE A COMPUTER	OTHER OCCUPATIONS
Richard MARTYN	Computer	1814–17	St Mabyn, Cornwall	Surveyor
Joshua MOORE	Computer	1798–1804	Cambridge	Uncertain
John PASCOE	Computer	1806–14	Devon	Surveyor
Reuben ROBBINS	Computer	1769–83	London	Mathematics teacher
Thomas SANDERSON	Computer	1807	London	Uncertain
Francis SIMMONDS	Computer	1805	Hampshire	Computer
Walter STEEL(E)	Computer	1769–76	Unknown	Computer
James STEPHENS	Computer	1776	Unknown	Ship's master
Michael TAYLOR	Computer	1779–98	London	Tablemaker
Philip TURNER	Computer	1805	London	Computer
John WALES	Computer	1779–96	Warmfield, Yorkshire	Uncertain
William WALES	Computer Comparer	1767–97	London	Astronomer Tablemaker Teacher
John WILLIAMS	Computer	1806	London	
George WITCHELL	Computer	1767–69	London and Portsmouth	Headmaster Computer
Benjamin WORKMAN	Computer	1815–16	London	Teacher

* These dates are the first and last *Nautical Almanac* editions that the computers and comparers worked on, and individuals may not have worked on every edition. Computation was often years in advance and in parallel, so the dates do not indicate when the work itself was done.

These stories about the *Nautical Almanac* computers show us two things about Maskelyne that contrast with the popular perception of him.

First, organizing the computation of the *Nautical Almanac* was a serious matter. Given the number of people involved and the variety of their location, experience and ability, it is a testament to Maskelyne's organizational skills and the robust systems he put in place to ensure that the *Nautical Almanac* was, as we would describe it today, fit for purpose.

Second, despite Maskelyne being portrayed in popular literature as a self-seeking academic astronomer with a less-than-personable style, the stories of his interaction with the *Nautical Almanac* computers reveal that he went to some lengths to provide stop-gap employment to mathematically inclined people, as well as providing long-term stable employment for those with families to support. However, as David Kinnebrook's story and that of other Royal Observatory assistants show, Maskelyne was a hard taskmaster who did not suffer those he considered fools gladly.

On Maskelyne's death in 1811, John Pond was appointed as the sixth Astronomer Royal, a post he held until 1835. Pond had good observational skills and updated the Observatory equipment, but he was a poor manager as well as suffering from recurrent ill health. This was coupled with his lack of interest in computational astronomy and the management of the *Nautical Almanac* fell to Thomas Brown, the replacement comparer to Malachy Hitchins. As mentioned above, Brown was more or less foisted onto Maskelyne by the Board of Longitude. He had no experience and was resented by more senior, competent and experienced computers such as Henry Andrews and Mary Edwards. When Pond took over, Edwards found that she was no longer being allocated the same amount of work, which meant her income was being cut by almost a third. She therefore petitioned the Board who agreed that she was a good and faithful worker and instructed Pond to pay her for twelve months' work rather than the eight he had allocated to her.[12] While this alleviated Mary's immediate financial distress it did not help to maintain her experience and competence with the *Nautical Almanac* computations.

By 1818, only seven years after Maskelyne's death, the quality of the *Nautical Almanac* tables had plummeted. Because of its advance preparation, it seems that standards began to fall from the 1815 edition, the first not to be compared by Malachy Hitchins. By 1818 things had become so bad that the issue was raised in Parliament. During a debate about the future of the Board of Longitude, the increase in errors in the *Nautical Almanac* was described, with the rider that the problem lay not with the scientific methods used to calculate the tables but in typographical errors – a statement which would seem to lay the blame squarely on the shoulders of the comparer, Thomas Brown, whose job it was to find such errors.[13] In response, Parliament appointed the well-known man of science and polymath, Thomas Young, as Superintendent of the *Nautical Almanac*, alongside his role as Foreign Secretary of the Royal Society. Young took over the task of day-to-day management of the *Nautical Almanac* work and retained both Maskelyne's calculating methods and the philosophy of using home-based workers to undertake the calculations. He retained the Reverend Thomas Brown (although it is not clear if he continued to be the official comparer), Eliza Edwards and three Cornish computers trained by Malachy Hitchins, namely Nicholas James, Richard Martyn and William Dunkin, as well as two other new computers.

Under Young's management the accuracy of the tables within the *Nautical Almanac* improved, but times had changed. In 1766, when it had been designed, its primary purpose was to support navigation with an emphasis on the calculation of longitude at sea. By the 1820s most navigators could afford a marine chronometer, so not everyone used the tables to their full extent. A new generation of astronomers wanted different, additional tables to further observing work using new advances in telescope technology. In addition, young mathematicians were developing new, more rigorous methods of mathematical table-making, not least to support the growing insurance and banking sectors, which made the techniques traditionally used to compute the *Nautical Almanac* look out of date.

The wider mathematical tensions in the Royal Society, the Board of Longitude, the Admiralty and the newly formed Astronomical Society were played out in the context of clamours for reform of the *Nautical Almanac*. When Thomas Young died in 1829, the modernizing mathematicians and astronomers got their way and it was decided that the *Nautical Almanac* would be remodelled and expanded, and the computational methods used updated. In 1831 the computers were given notice that, once they had completed the work for the 1833 edition, their services would no longer be required. The new *Nautical Almanac* would be computed in an office in London, using different techniques. Only one of the existing computers, William Dunkin, who was in his fifties by this time, decided to make the move to an office in central London. While the *Nautical Almanac* soon regained its good reputation, for Dunkin it was not a good move. His son recalled: 'I have often heard him express a real regret at the loss of his semi-independent position at Truro, in exchange for the daily sedentary confinement to an office-desk for a stated number of hours in the company of colleagues all junior to himself in ages and in habits.'[14]

Maskelyne may well have approved of the changes made to the content of the *Nautical Almanac* (although his priority was always that of the sailor over the astronomer) but one cannot help feeling he would have been dismayed at the treatment of the individuals who had served it for so many years.

FURTHER READING

My research on the eighteenth-century computers of the *Nautical Almanac* has been published in 'Providing Longitude for all', *Journal for Maritime Research* (Sept 2002) and 'Tabulating the

Heavens: Computing the Nautical Almanac in 18th–Century England', *IEEE Annals of the History of Computing*, 23 (2003), pp. 48–61.

There are also articles which focus on the individual *Nautical Almanac* computers, for example my 'Mary Edwards: Computing for a living in 18th-Century England', *IEEE Annals of the History of Computing*, 23 (2003), pp. 9–15 and 'Henry Andrews (1744–1820): An Astronomical Calculator from Royston', *Herts Past & Present*, 3/2 (2003), pp. 21–27. Other biographical articles feature *Nautical Almanac* computers where their computing is a very minor part of their life's work, such as Wayne Orchiston and Derek Howse's 'From transit of Venus to teaching navigation: The work of William Wales', *Journal of the Institute of Navigation*, 53 (2000), pp. 156–166 and Lynn B. Glyn's 'Israel Lyons: A short but starry career. The life of an eighteenth-century Jewish botanist and astronomer', *Notes and Records of the Royal Society London*, 56 (2002), pp. 275–305. David Kinnebrook's experiences have been put into context in Simon Schaffer's 'Astronomers mark time: discipline and the personal equation', *Science in Context*, 2 (1988), pp. 9–23.

For information on the life of related workers in the eighteenth century also see my 'Astronomical Labourers: Maskelyne's Assistants at the Royal Observatory, Greenwich, 1765–1811', *Notes and Records of the Royal Society*, 57 (2003), pp. 285–98.

There were other groups of human computers in the eighteenth and nineteenth centuries. These are described well by David Grier in *When Computers were Human* (Princeton: Princeton University Press, 2005). For an English perspective see my 'Human computers in eighteenth- and nineteenth-century Britain', Chapter 4 in Eleanor Robson and Jacqueline Stedall's *The Oxford Handbook of the History of Mathematics* (Oxford: Oxford University Press, 2008). Life at the nineteenth-century Royal Observatory for junior assistants and computers is described in Edwin Dunkin's *A Far off Vision: A Cornishman at Greenwich Observatory*, eds. Peter Hingley and Tamsin Daniel (Truro: Royal Institution of Cornwall, 1999). Nineteenth-century tide-table computers are discussed in Chapter 6 of Michael S.

Reidy's *Tides of History: Ocean Science and Her Majesty's Navy* (Chicago: University of Chicago Press, 2008).

While not strictly about computers, Allan Chapman's *The Victorian Amateur Astronomer* (Chichester: John Wiley and Sons, 1998) gives a good flavour of both well-known and less well-known astronomers in the nineteenth century, with insights into how some of them earned their living.

The tensions in astronomy and mathematics in the 1820s are well described by William Ashworth's 'The calculating eye: Baily, Herschel, Babbage and the business of astronomy', *British Journal for the History of Science*, 27 (1994) pp. 409–41 and David Philip Miller 'Between hostile camps: Sir Humphrey Davy's Presidency of the Royal Society of London, 1820–1827', *British Journal for the History of Science*, 16 (1983), pp. 1–47.

CASE STUDY D

MASKELYNE AND THE MARINE TIMEKEEPER

Although, as mentioned in Chapter 3, it was not obvious that the Royal Observatory should have become the place in which marine timekeepers were tested, this was one of the new duties that arrived during Maskelyne's tenure there. Like the publication of the *Nautical Almanac* it was to be a long-term element of work at the Observatory. In the few decades after Maskelyne's death it was increasingly formalized, with logs of chronometer testing, rating and distribution being kept, prize trials for new timekeepers run and (from 1833) time checks for ships in the Thames and expanding London docks provided by the daily signals of a time-ball atop Flamsteed House.

When Maskelyne entered that building, however, there were but few marine timekeepers in existence. Harrison's watch (H4) was the first to undergo a Greenwich trial in 1766–7. This revealed, if nothing else, the care with which timekeepers had to be adjusted and rated in order to perform usefully. Maskelyne in part took on this job because he had become the Astronomer Royal and, therefore, the chief arbiter of matters scientific and technical on the Board of Longitude. However, it also followed naturally from the fact that he had been involved with trialling longitude methods in the field, rather than the observatory, during his voyages to St Helena in 1761–2 and Barbados in 1763–4.

While Maskelyne had been appointed as a transit of Venus observer by the Royal Society in 1761, for his second expedition he was under the instructions of the Board of Longitude. He was there to help carry out the required West Indies voyage dictated by the original Longitude Act, during which the ship making use of the method submitted should travel to a West Indies port 'without Losing their Longitude beyond the Limits' of a degree or half a degree.[1] However, as had not been foreseen in 1714, this 1763 trial was not just testing one method but three.

As 'methods' each was considerably different. Actually under trial were Mayer's lunar tables (tested through use of the lunar-distance method), a marine chair designed by Christopher Irwin (tested through its ability to allow someone to observe Jupiter's satellites) and Harrison's sea watch. Maskelyne and his assistant Charles Green, the assistant to Bradley and Bliss at the Royal Observatory, were required to carry out, at sea, the observations needed to test the first two methods and then to establish, on land, the longitude of Barbados by observing Jupiter's satellites. The results achieved by all three methods would be checked against this newly established figure.

Harrison's H4 travelled separately, in the care of William Harrison, meaning that Maskelyne had not yet become familiar with its use. It is certainly the case that he was much more familiar with astronomical methods, and was fully persuaded of their potential and usefulness. Nevertheless, there is no evidence to support the accusations that the Harrisons were subsequently to raise about his bias. Although Maskelyne may have seen a source of income through the publication of further books like *The British Mariner's Guide*, he was neither up for a reward from the Board of Longitude nor would he have felt that the success of Harrison's timekeeper would have rendered an almanac redundant. The two approaches being complementary, he consistently saw the advantage of making both available.

Irwin's chair was dismissed but the results of the Barbados trial were good for the other two methods. While Mayer's tables led to a less accurate result than the watch, the lunar-distance method had the advantage in that it could immediately start to be implemented through training and publication of the *Nautical Almanac*. It was also anticipated that further work would improve the accuracy of the lunar theory, and thus of the method. Another point in favour of astronomical methods was that each determination of longitude during a voyage was made afresh, while timekeepers might produce cumulative error.

However, weighing the lengthy observations and calculations associated with lunars against the comparably easy use of a timekeeper, it was obvious to all that the ideal solution was to have one of these on every vessel, with its going checked periodically by astronomy.

The question for the Board was how to turn Harrison's unique and expensive watch into a 'method' that could be applied on more than one ship. The Commissioners chose to interpret the method that Harrison had put under trial not as the watch itself but, rather, the means by which such watches could be made. Their recommendation, on 9 February 1765, was that the money given to Harrison so far would be made up to £10,000 'upon his producing his Timekeeper to certain persons to be named by the Board & discovering to them, upon oath, the principles & manner of making the same'. Another £10,000 would be granted if the Board could be convinced that 'his method will be of common & general Utility in finding the Longitude at Sea'.[2] It was later decided that this would only be proved if further timekeepers were made according to Harrison's method and trialled successfully.

Harrison was fiercely opposed to the Board's interpretation of the original Longitude Act and to its having ensured the passing of a new Act, which shifted the goalposts dramatically. He always maintained that he had met the terms of the original legislation and deserved the £20,000. The Board, however, felt obliged to find a way to extract useful information from him and to invest in the development of the complementary astronomical methods.

It was through this process that Maskelyne came to be ever-more closely involved in the testing and adjudicating of timekeepers. He was to oversee the 'discovery' of Harrison's watch before witnesses, to edit the published description of the watch's mechanism and to subject all of Harrison's timekeepers to further trials at the Royal Observatory. As a result of this and of his earlier involvement in testing the lunar-distance method at sea, the Harrisons came to single him out as the specific enemy within the general foe that was the Board of Longitude. He was attacked in print, especially in the 1767 *Remarks on a Pamphlet lately published by the Rev. Mr. Maskelyne.*

In a letter of 1773 to Lord Sandwich, then First Lord of the Admiralty, Maskelyne mentioned having considered publishing a response to Harrison's pamphlet, but that he had decided against it. While his express reason was that 'thinking such abuse thrown out without probability or proof required no refutation', he was perhaps also satisfied that most of his Board colleagues backed him.[3] As Derek Howse wrote, 'both contemporary and present-day accounts refute most of the accusations' that Harrison brought against Maskelyne.[4]

Despite Harrison's complaints and protestations, the Board more or less succeeded in getting what it required from him. However, the information was far from perfect and both those watchmakers who tried to make sense of the published account and those who were present at the 'discovery' found it inadequate to make another such timekeeper. Somewhat more successful was the attempt to show that a copy of Harrison's watch could be made by another craftsman. The Board selected Larcum Kendall, who had probably assisted with making the original watch, to produce the replica.

It took Kendall two years to make his copy, now known as K1, for which he was paid £450 by the Board. Kendall's view, however, was that it represented the wrong approach. If the Board wanted something that could, ultimately, be made quickly and cheaply enough to be of widespread use, then the design and its use would have to be simplified. The Board agreed to let Kendall try his new ideas and his subsequent timekeepers K2 and K3 were the result, costing £200 and £100 respectively. While H4 never went to sea again after the Barbados trial, all of Kendall's timekeepers had an early history of interesting and sometimes dramatic voyages, such as when K2 was taken by the *Bounty* mutineers to Pitcairn Island.

Neither Kendall nor Harrison – who, under duress, was working on his second sea watch (H5) – were the sole individuals working on timekeepers, nor was the nascent 'marine chronometer' simply an outcome of their work. In France, Pierre Le Roy and Ferdinand Berthoud were coming up with a range of designs and machines, some of which incorporated important innovations and tested well at sea. In London, John Arnold, to whom Maskelyne had made sure to send a copy of the description of Harrison's watch mechanism, was interesting the Board of Longitude with promises of good performance for significantly cheaper prices. It was through the work of these men, later joined by Thomas Earnshaw, that the standard chronometer – a precision instrument available in sufficient numbers at affordable prices – was developed.

Each of the early timekeepers presented to the Board was tested by Maskelyne at Greenwich. His trials showed that Kendall deserved his payments and that Arnold's and Earnshaw's designs could perform exceptionally well. Where the Board had presented a reward or paid for production, the timekeeper came into its possession, joining a slowly expanding collection of instruments that Maskelyne would ensure were put to work overseas. As discussed in Case study E, it was through his initiative that the late-eighteenth-century voyages of scientific exploration were able to make use of timekeepers at sea.

The evidence therefore suggests that, rather than being hostile to the timekeeping method of determining longitude, Maskelyne was deeply interested in its progress, hastening it along where he could. It is also interesting to note that he was sufficiently open-minded to spot the potential of using timekeepers in an attempt to measure the longitude difference between Greenwich and Paris. At the same time that General Roy, the leader of the national Trigonometrical Survey (later known as the Ordnance Survey), was attempting to carry out a geodetic connection between England and France, Maskelyne sent Joseph Lindley, his assistant, off by coach and boat to Paris with eight Arnold chronometers and watches. Although Maskelyne never made this result official, he clearly valued it: it also turned out to be better than that achieved by Roy.

It is true that Maskelyne was criticized by watchmakers other than Harrison. Thomas Mudge asserted that Maskelyne had a pecuniary interest in the lunar-distance method, and that he was biased against the longitude timekeepers and the 'mechanics' who made them. Mudge also felt that the subsequent Longitude Acts of the 1770s had made it easier for makers of astronomical tables than for watchmakers to receive rewards. Later, the son and heir of John Arnold, whom Maskelyne had certainly supported at various points, felt aggrieved that Maskelyne had taken Earnshaw's side in the dispute over who had invented the detached chronometer escapement.

Maskelyne also occasionally offended or received criticism from others who were not watchmakers. In the end, such accusations were part and parcel of his being the one sitting in judgement on all ideas and devices, as well as sometimes being the product of particular groupings and loyalties. Much of what was under dispute was, in any case, still in flux. For example, Maskelyne was criticized by Harrison and Mudge for the manner in which he rated timekeepers that were under trial, largely because it made their timekeepers' performance less impressive than their preferred method would have done: yet Maskelyne's approach was probably the best then available and he was at least consistent in the way he applied it.[5]

It seems clear that he was exacting in his assessments and unwilling to back projects and proposals, especially with public money, if he could not be sure that they deserved it. He was undoubtedly most sympathetic to those who were similar to himself, in terms of manner and interest in things mathematical, and could be dismissive of those whom he felt lacked the capacities necessary to a particular task, emphatically including practical and mechanical skills. He was a practical man and a tinkerer himself, who valued the makers of the instruments and clocks used within his own observatory and on expeditions, of which he also had practical field experience.

5

MASKELYNE'S TIME

Rory McEvoy

The type of work carried out at the Royal Observatory, Greenwich, up to and during Maskelyne's tenure as Astronomer Royal was not concerned with cosmology but providing the astronomical data for mariners to find their position at sea reliably using the stars, planets and Moon as markers. As earlier described, positional astronomy involves observing the apparent east-west motion of the sky and timing the precise moment that a given star or celestial object passes directly overhead and its angular height. In simple terms it requires two basic instruments: a clock and a telescope.

The telescope, known as the transit instrument, has to be exactly fixed in a north-south plane (that is, on a meridian) so that the observer will always see the stars passing horizontally across the field of view with his meridian at its vertical centre. The quality of the observations depends on the transit clock's stability and accuracy. The terminology for this type of clock is used loosely: in observatories it is generally known as an astronomical clock but in horological workshops as a regulator.

This chapter will identify the principal astronomical clocks and watches used by Maskelyne when the Royal Observatory was, for the first time, a centre for testing the emerging technology of the sea-going watch, as well as looking at his role in commissioning improvements and trialling new astronomical clocks destined for observatories elsewhere. Where possible, it will explore his relationship with their makers and, in so doing, further examine and seek to dispel the peculiar (and generally quite recent) popular notion that Maskelyne was somehow antagonistic towards horologists.

THE ASTRONOMICAL CLOCK

Throughout Maskelyne's tenure as Astronomer Royal he used the same astronomical clock, and on the Greenwich meridian defined (since 1750) by Bradley's transit instrument. The clock, which we met briefly in Chapter 3, was the third of its type purchased for Greenwich from the London-based instrument maker, George Graham, and is now known as Graham 3. During Maskelyne's forty-six-year term, Graham 3 was frequently altered and improved and therefore plays a central role in this investigation of his relationship with the horological trade.

So what constitutes an astronomical clock? The key elements are defined by Maskelyne in his list of essential instruments for the 1761 transit of Venus expedition to St Helena, proposed two years after he was honoured with Fellowship of the Royal Society.[1]

Firstly, 'The clock must have an [*sic*] hand to distinguish seconds'. In order to make meaningful observations, the clock had to show seconds clearly and audibly. The observer would glance at the clock dial to read the seconds and return attention to the telescope, counting the seconds during the observation by listening to the ticking of the clock. This was known as the 'eye-and-ear' method. Almost all astronomical clocks employ a pendulum of around thirty-nine inches long to tick once per second.

Secondly, it 'should be one of those called regulators which are so made as to lose no time while they are wound up'. A weight-driven clock will inevitably lose time if it is not fitted with a device that provides a temporary driving force during winding. The technical general term for such a device is 'maintaining power'.

Graham 3 was originally fitted with bolt-and-shutter maintaining power, which is simple and effective. When the clock is due to be wound, the operator slides a pin located at the side of the dial, which engages a sprung lever into the teeth of the centre wheel, forcing it to maintain its normal clockwise motion. At the same time a shutter is drawn back, revealing the small winding square. This system provides temporary power for around two to three minutes, which is ample time to wind the clock, while the presence of the shutter ensures that the clock cannot be wound unless the maintaining power is activated.

The third requirement is to have a 'pendulum so made as not to be affected by heat or cold'. The clock's timekeeping is governed by the length of the pendulum and the pendulum rod is made of brass or steel, so the clock's rate will slow down when it is hot, owing to expansion of the rod, and vice versa when cold. Clocks were therefore commonly fitted with a grid-iron pendulum to overcome the ill effects of temperature change on timekeeping. The grid-iron is an ingenious device, invented around 1725 by John Harrison, which harnesses the unequal expansion rates in steel and brass by employing five steel and four brass rods arranged in opposition, so that when they expand and contract the overall length of the pendulum remains constant.

Precision clockmaker, John Shelton (*c.*1698–*c.*1775), whose business card described him as 'operator to the late Mr Graham', was chosen to supply the astronomical clock for Maskelyne's part in the St Helena expedition. This was bought for the purpose by the Royal Society. If it was Maskelyne who made the choice, then his decision was probably influenced by his earlier experience working with James Bradley.

Maskelyne tells us that he assisted Bradley with mathematical calculations to confirm the latitude of Greenwich 'after he [Bradley] became possessed of the new instruments in 1750' (i.e., those then supplied to the Royal Observatory) and that Bradley had acquainted him with the method of observation used to find latitude and degrees of atmospheric refraction.[2] Atmospheric refraction causes the beautiful sunsets which most of us enjoy and is a lensing effect produced by the curvature of the Earth's surface and the layers of gases that hug it: however, it makes accurate positional observation of stars increasingly difficult the closer they are to the horizon. This particularly affects transit instruments when making low-angle observations rather than high ones: by contrast the zenith sector is vertically orientated and has minimal angular adjustment, and therefore views the least distorted part of the sky.

Bradley had used a zenith sector made by Graham to great effect to discover the aberration of light and the Earth's nutation (slight wobbling on its axis) which were arguably the two most important observational discoveries in astronomy during the first half of the eighteenth century (Fig. 1).

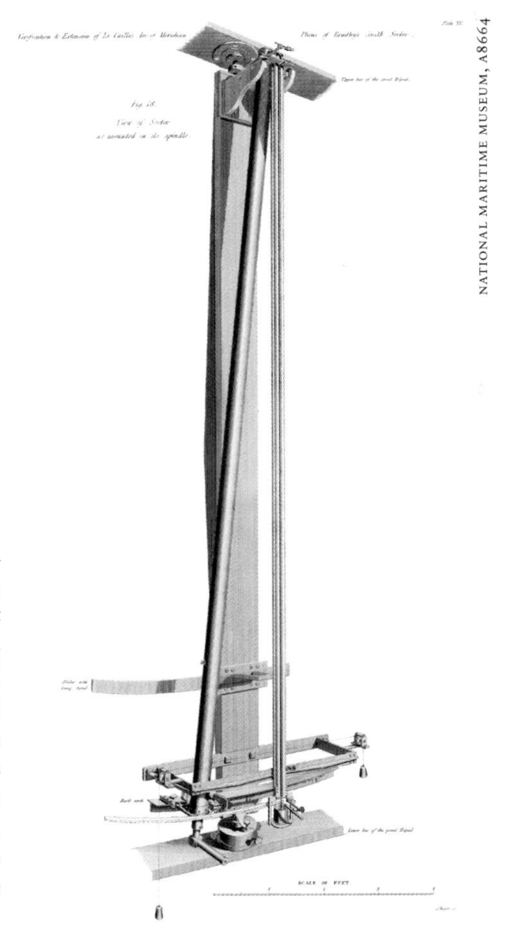

Fig. 1: Bradley's zenith sector telescope, made by George Graham, is an improved version of one used at Kew in the 1720s by Samuel Molyneux (1679–1728) and James Bradley. It was the principal instrument used to discover the aberration of light and the Earth's nutation. This engraving is from an 1866 work by Thomas Maclear, recording his later use of the instrument for survey work in the Cape of Good Hope.

Graham's part in them is credited by Bradley thus:

> For I am sensible that if my own Endeavours have, in any respect, been Effectual to the Advancement of Astronomy; it has principally been owing to the Advice and Assistance given me by our worthy Member [of the Royal Society] Mr. George Graham; whose complete and practical Knowledge of the Uses of Astronomical Instruments, enable him to contrive and execute them in the most perfect manner.[3]

Exactly when Maskelyne assisted Bradley is not known but it was probably before he graduated from Cambridge. The association certainly served Maskelyne well, for in 1758 he in turn was elected a Fellow of the Royal Society, Bradley being among his six proposers.

John Shelton probably began to work for Graham immediately after serving his apprenticeship and made astronomical clocks for him from his own workshop near Fleet Street[4]. After Graham's death in 1751, Shelton continued producing such clocks on his own account from a workshop in Shoe Lane, on very much the same pattern that he had done under Graham. If one compares the movement of one of Shelton's transit-of-Venus clocks to an early example signed by Graham, there will be minor peripheral differences but their layout, manner of construction and dimensions will be practically identical.

Another astronomical clock, made by John Ellicott (1706–72), was purchased for the second party observing the 1761 transit of Venus, which headed for Bencoolen but ultimately observed it from the Cape of Good Hope. The ambitious Ellicott became a Fellow of the Royal Society shortly after submitting a paper (in 1735) on the measurement of expansion in metals caused by raised temperature.[5] For the 1761 expeditions he offered to lend the Society an equal-altitude instrument, perhaps as a sweetener to secure the prestigious commission to supply an astronomical clock.

As the St Helena mission is dealt with elsewhere in this volume, it is unnecessary to say much beyond noting that Maskelyne used the clock made by the 'diligent and ingenious artist Mr John Shelton' to measure local gravity, as part of continuing study of the shape of the Earth.[6] His experimentation with the clock on St Helena is indicative of Bradley's influence, for in 1733 Bradley encouraged readers of the *Philosophical Transactions* to conduct such experiments when abroad in his write-up of Colin Campbell and George Graham's investigation at Black River, Jamaica.[7]

Campbell had used a plain pendulum of fixed length in conjunction with a thermometer to determine the effect of temperature on the rate of the clock. Unlike Campbell, Maskelyne's clock was furnished with a temperature-compensated pendulum and so he was unable to follow Campbell's method exactly, but he also used a thermometer as a safeguard. Maskelyne noted finding gravity reduced at St Helena and suggested that Earth's structure and density needed to be better understood before any conclusions could be drawn. In 1774 he continued this study at Schiehallion, again using a Shelton clock purchased by the Royal Society for the transit of Venus expeditions of the 1760s, although apparently not the one he had taken to St Helena.[8]

BARBADOS

Maskelyne used the same Shelton astronomical clock again in 1763–4, when he was appointed to sail to Barbados ahead of John Harrison's marine watch, H4, to determine the longitude of that island. In a letter to his brother, Edmund, dated 29 December 1763, Maskelyne briefly described the outward voyage; also how he had trialled the lunar-distance method on it with the same success as on his earlier passages to and from St Helena, and how he had fixed the longitude of Barbados to within half a degree. He also mentioned that he had tested Christopher Irwin's chair – a gimballed and weighted observational seat, intended to counteract a ship's motion – and proved it to be 'a mere bauble'.[9]

It has to be said that this flash of humour is a rare direct glimpse of the human being in Maskelyne. He certainly had reason to be happy, if only for his success in further vindicating the value of the lunar-distance method of finding longitude and, as a man known at least occasionally to be 'convivial', the good hospitality and balmy surroundings he experienced in Barbados no doubt helped. We cannot be sure if he imagined, while on the voyage, how soon the position of Astronomer Royal at Greenwich would again fall vacant but he was also undoubtedly aware that the still-recently installed incumbent, Nathaniel Bliss, was sixty-three (in November 1763) to his own thirty-one. It is inconceivable that such a professionally focused man did not have possible future career openings in mind, in whatever timescale, and he had every reason to believe that the results of the Barbados voyage could only further recommend him for advancement when suitable opportunity offered. Events were to prove him right when Bliss died very suddenly, still sixty-three, in September 1764.

We know that William Harrison was not happy to hear of Maskelyne's success with the lunar-distance method, which he undoubtedly considered a considerable threat to his father's work. He questioned Maskelyne's impartiality on the performance of H4 in its Barbados trial and accused him of botching the observations required to judge it. Thus began the emotive struggle of the Harrisons to secure the full reward offered by the 1714 Longitude Act.

The pertinent criticism of Maskelyne is of his handling of the timekeepers. On 23 May 1766, he collected the three large ones (now called H1, H2 and H3) from Harrison's home in Red Lion Square. He was criticized by Harrison for using an un-sprung cart and there was an incident in which one of them was damaged by a dropped pair of pincers, which 'broke some of the movements therein at which Mr Maskelyne's hand was then on the machine'.[10] There was further upset over the testing of H4 at the Royal Observatory in 1766, after which Maskelyne concluded that 'Mr Harrison's watch cannot be depended upon to keep the longitude within a degree in a West India voyage of six weeks'.[11] Here the problem was not so much with Maskelyne's handling of the watch but more the lack of preparation of it for trial. The Board did not appreciate that this timekeeper required careful cleaning and adjustment prior to its use and expected it to perform well when required. Even setting aside the Harrisons' opinions on the matter, they were unquestionably the best qualified to do this, but had no involvement in the process prior to testing against Graham 3 and the transit instrument at Greenwich.

As the John Harrison story is amply covered in other volumes, there is little benefit in pursuing it here but it is worth mentioning his interest in pendulum clocks. He published a peculiar book about clock-making in 1775, a year before his death, in which he made an extraordinary claim for his radical design of pendulum clock – accuracy to within one second in one hundred days. (Graham 3, one of the most accurate clocks of the time, would have been susceptible to changes in barometric pressure and the consistency of its oil, so would have been accurate to around one second per week.) In the book, he takes a passing shot at Maskelyne by saying 'I once thought of giving a clock to the observatory at Greenwich, but my bad usage proved too tedious for that'.[12] The unproven claim is unlikely to have troubled Maskelyne much, but it would have highlighted scope for improvement and may have inadvertently helped prompt him to turn to other makers of marine timekeepers for improvements to Graham 3.

ASTRONOMER ROYAL

Maskelyne's appointment as Astronomer Royal, following the death of Nathaniel Bliss (1700–64), represents a significant turning point in the Royal Observatory's history. The cumulative efforts of Maskelyne's predecessors there, and others (including himself up to 1765), meant that the founding goal of the institution, to 'find longitude' by astronomical means, had been realized. From very shortly after he moved in, it henceforth became an establishment that broadcast astronomical information in the form of the *Nautical Almanac*. The gathering of astronomical data was an ongoing part of business at the Observatory and Maskelyne aimed to improve the quality of that information.

A related new responsibility for the Astronomer Royal was the testing of new marine timekeepers. This role put Maskelyne in direct regular contact with some of the most talented horologists of the time and this circumstance often proved beneficial to him as well as to the makers.

Advance preparation for Cook's second Pacific voyage of discovery (1772–5), which was pivotal in confirming the global practicality of both the marine timekeeper and the lunar-distance method for determining longitude at sea, put Maskelyne in charge of testing the timekeepers to be used on it. Both Kendall's copy of H4 (now known as K1), and a timekeeper by John Arnold (1736–99), probably Arnold No. 3, were tested at the Royal Observatory before being embarked.

JOHN ARNOLD

The fact that Arnold was commissioned in July 1771 to make improvements to Graham 3 is no coincidence and strongly suggests that, once discussion of the timekeeper for Cook was over, Maskelyne and Arnold's conversation moved on to potential improvements of the astronomical clock.

Arnold improved the clock by fitting jewel pallet nibs to the dead-beat escapement to reduce friction. Jewels were not new to watchmaking – they had been used as pivot holes for the train wheels since the early 1700s – but furnishing a clock escapement with jewels was quite a radical new step and met with criticism among Arnold's peers. Clearly the condemnation was unwarranted since Maskelyne was pleased with the results and commissioned Arnold to make two new clocks, also with jewelled escapements. These clocks are numbered Arnold 1 and 2, which indicates that commercial production of astronomical clocks or regulators was a new venture for him, initiated by Maskelyne's encouragement.

Used in the Observatory's Quadrant Room and east dome, both these clocks survive but were sold in the 1930s, each to separate clock dealers, one of whom in a letter of negotiation cheekily excused his low offer by dismissing them as 'derelict plant'.[13] It is very interesting and noteworthy that the movements of both closely follow the design of Graham 3, featuring tall plates with tapering shoulders (which are also fully jewelled) to give extra rigidity to the escapement.

Arnold's introduction to Maskelyne was timely and fortuitous, for Maskelyne advised on instrumentation for a new observatory being built by the Elector Palatine, Charles IV Theodore, in Mannheim. Arnold was recommended and duly commissioned to make an astronomical clock, which was tested to Maskelyne's satisfaction at the Royal Observatory for six months before its dispatch to the Holy Roman Empire.

The Mannheim commission proved lucrative for Arnold. The Elector's astronomer, Christian Mayer, was very pleased with the clock and wrote a lengthy letter generously commending its quality. Arnold was not slow to take advantage of a good business opportunity: he had the letter translated into English and published it as a pamphlet.

The translator mentions in the preface the advantage gained by Arnold's method of suspending the pendulum bob from the centre of oscillation rather than its base, to eliminate changes in rate caused by the bob itself expanding upwards. The centre of oscillation differs slightly from the bob's physical centre and, interestingly, it was Nevil Maskelyne who supplied the mathematical data to suspend the bob to the best advantage. This appears to be the only known example of Maskelyne contributing directly to the improvement of astronomical clocks and should not be confused with his unrelated work on the plumb-line of zenith sectors.

As a result of the success of the ruby pallets in Graham 3, Arnold fitted them to the two clocks taken by William Bayly and William Wales on Captain Cook's second voyage of discovery; the bill, submitted to the Board of Longitude in March 1772, came to £27 3s. There is evidence that around this time many existing clocks and regulators were sent to Arnold to have their escapements jewelled, probably as a direct result of the published letter of approval from Mayer.

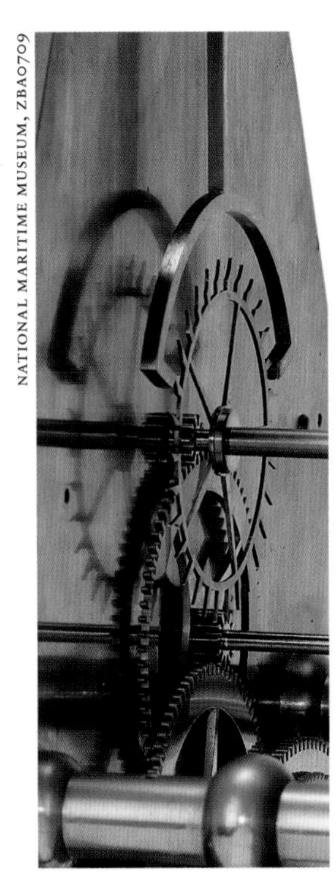

Evidently Maskelyne felt that further improvement could be made to Graham 3 and he called upon Arnold's services again. In February 1779, Arnold returned Graham 3 to the Observatory after a fairly radical overhaul. He replaced the brass escape wheel with one of steel and removed the bolt-and-shutter maintaining power in favour of Harrison's version of the same, which automatically engages when the winding key is turned, thus negating the need for a shutter (Fig. 2).

Fig. 2: A detail from Graham 3, showing the jewelled escapement and steel wheel.

Again, the treatment to Graham 3 coincided with the submission of another Arnold timekeeper, watch No. 36, which was the first of its type to have temperature compensation fitted to the balance (Fig. 3).

The watch was trialled at the Royal Observatory for thirteen months from February 1779 and its performance was unprecedented.[14] Rather than pursue a reward, Arnold saw a greater business opportunity and published the outstanding performance of No. 36 in a pamphlet.

This pamphlet marks a milestone in the history of the marine timekeeper as it defines Arnold's watch as a 'chronometer', a term which quickly became standard. Although the word, which is derived from Ancient Greek, had been used in the past to describe various timekeeping devices, this is the first specific use of it in the latter half of the eighteenth century to describe a sea-going watch.

Fig. 3: Arnold No. 36, the subject of Arnold's 1780 pamphlet in which the term chronometer was formally used for the first time.

NATIONAL MARITIME MUSEUM, ZBA1227

WILLIAM COOMBE

John Arnold was an established watchmaker and, as has been shown, his interaction with Maskelyne proved to be very good for business, encouraging him to diversify beyond the making of watches and chronometers. The next horologist to come into contact with Maskelyne at Greenwich was altogether a different character. William Coombe, a previously unknown watchmaker, also submitted a timekeeper for trial to the Observatory at the same time as Arnold No. 36 was being tested. Coombe does not appear in the records of the Worshipful Company of Clockmakers as having served an apprenticeship. He appears likely to have been a 'gentleman horologist', indulging in watchmaking for his own pleasure rather than as a business concern.

The two men appear to have got on rather well: Maskelyne comes across as patient, supportive and perhaps even a distraction to Coombe. We know very little about Coombe other than from a watch and an astronomical clock that belonged to

Maskelyne and a few letters among the papers of the Board of Longitude. The marine timekeeper trialled in 1779 is not known to have survived. From the minutes of the Board we know that it performed well enough to earn a recommendation from the Astronomer Royal, resulting in financial encouragement of £200.

Before continuing with William Coombe's story, it should be said that Maskelyne was testing two other timekeepers at the same time. These were a pair by the name of Thomas Mudge known as 'Green' and 'Blue'. They did not perform so well and did not qualify for any reward. Thomas Mudge junior felt that his father had been wronged and he petitioned for a reward to the Board of Longitude and Parliament, alleging in print that Maskelyne had 'very *liberally* declared, *they had given the Mechanics a bone to pick that would crack their teeth*' when commenting on the 1774 revision of the Longitude Act. Statements like this have, of course, served to propagate the idea that Maskelyne was indeed anti-horologist, but by this stage it seems very absurd that he would have deliberately obstructed a longitude reward simply because he preferred the astronomical method.[15]

As with Arnold before, Coombe was recommended by Maskelyne to make an astronomical clock for the Imperial and Royal Museum of Physics and Natural History in Florence. Not only did Maskelyne recommend Coombe; he paid him funds in advance. Maskelyne's account books show that Coombe was paid the final £21 for the clock in July 1783, bringing its total cost to £63. As with Arnold's before, it was brought to Greenwich for trial before dispatch to Florence. We learn from Coombe, in a letter written in 1791, that it went well there for four months but then needed to be taken back for adjustment.

It is interesting that, despite his clock's apparent poor performance, Coombe's relationship with Maskelyne remained good. Maskelyne notes taking delivery of an unusual short-interval timer made by Coombe in July 1786 (Fig. 4).

He kept this for himself and ordered another from Larcum Kendall (1721–95), well known to him as the Board of Longitude's officially commissioned copier of H4.

Coombe's timer is a stop-watch with a gilt-brass, open-faced

NATIONAL MARITIME MUSEUM, ZBA4670

Fig. 4: Maskelyne's short-interval timer, which is capable of measuring to within one tenth of a second.

case and is designed to beat 36,000 times per hour, which means that it can indicate time to tenths of seconds. The signed enamel dial has three separate dials showing tenths of seconds, whole seconds and minutes. The dials are arranged in the same manner as on Coombe's 1781 regulator and bear passing resemblance to the dial of Kendall's K3 timekeeper, made in 1774. The watch has a stop/start lever beside the dial which can only be accessed with the front bezel open.

For an astronomer, a watch such as this could have many different uses; for example, comparing the rate of journeyman and other pendulum clocks with the main transit clock as well as timing observations such as equal altitudes, for establishing local time, or lunar distances. Maskelyne was already able to judge the timing of transits to an accuracy of one-tenth of a second using the 'eye-and-ear' method, which had been devised by his predecessor, James Bradley, and the watch may have played a part in evaluating the accuracy of his observations.[16]

It seems as though Coombe was not financially secure for, five months after Maskelyne's payment of £21 for the astronomical clock intended for the Grand Duke of Tuscany in Florence, he wrote to the Board of Longitude requesting further financial assistance. The request was turned down and Coombe was told that he would have to provide evidence of making further horological improvements before any reward could be considered.

Maskelyne, it appears, was sympathetic to Coombe and may have encouraged the Royal Society to employ him to repair and maintain one of the Shelton astronomical clocks, its deal frame and a journeyman clock. These were collected from the Royal Society and delivered back to Maskelyne at Greenwich when ready. Coombe's bill also shows work for the Royal Observatory, which is crossed out and the time charged amended, suggesting that Maskelyne had organized the work and employed him separately to clean the Observatory's transit-of-Venus clock.[17] The bill also mentions cleaning a watch 'with seconds of unusual construction', possibly the Ellicott watch which had been purchased by the Board from Maskelyne a few years earlier.

Coombe's 1783 letter requesting assistance from the Board is particularly interesting as it indicates that he was quite involved with Maskelyne. He tells us that he has been researching different types of pendulum as well as investigating improvements to Hadley's quadrant.[18] We also learn that he had produced a large marine watch that could be carried in a coat pocket or box, but there is an air of desperation in his writing which suggests that he may have been lucky with the performance of this timekeeper and, when he tried to improve it, got out of his depth.

This impression is reinforced by a letter written to the Board in 1799, when he summarizes his efforts over the previous twenty-one years and estimates the costs of construction and 'destruction of watch work' to be in the region of £700. He ends imploring the Board 'to set me on a level with the world again, as I was at first'.[19] Needless to say the Board remained firm in its stance and sent no reward to Coombe. What happened to him thereafter is unknown.

It is perhaps worth noting that Kendall's commission to make an astronomical clock for the Grand Duke of Tuscany after Coombe's failed to perform adequately also coincided with further improvement to Graham 3. In 1789 Kendall replaced the pendulum bob and altered the pendulum to allow for finer regulation of the clock. No further alterations were made to Graham 3 for another four years, until the gifted watchmaker, Thomas Earnshaw, came to Maskelyne's attention through introduction to him by Mr Thomas Noyes, a farmer from Eltham.[20]

THOMAS EARNSHAW

Earnshaw (1749–1829) was ambitious and hard-working, intent on turning around his fortune from that of back-room watch finisher to successful watchmaker in his own right. Like Arnold, he played a significant part in the development of the marine chronometer but their two characters could not have been more different. Arnold was an excellent businessman, knowing the value of contacts and powerful allies. Earnshaw, by contrast, was more terrier-like, confident in his own abilities and not afraid to upset people when it came to fighting his corner. In fact, Earnshaw managed to make himself an enemy to practically the whole horological trade in vigorous defence of his claim to have invented the spring detent and his protestation that Arnold had plagiarized the idea.

In simple terms, the detent is the heart of the chronometer escapement and the spring detent was a significant improvement of the pivoted detent. The weakness lies in the pivots, which require oil, and so the timekeeping will be affected by changes in the oil's lubricating properties. Oils available at the time were liable to change their properties quickly and the spring detent eliminates part of this problem.

Earnshaw conceived the idea of the spring-detent sometime in 1780 but could not to afford to file a patent under his own name, so had Thomas Wright take out the patent on his behalf in 1783. Unfortunately for Earnshaw, Arnold had already taken out a patent the previous year for a slightly different form of spring-detent escapement. It seems quite likely that Earnshaw had the original idea and that Arnold heard enough about this to be able to create his own design. However, without controversy over the invention historians would be much worse off, for Earnshaw's agitations stirred a lot of heated discussion, particularly after Arnold's death in 1799.

Earnshaw continued to petition the Board of Longitude for reward and it was not long before his 'irritable habit' caused him to cross swords with the powerful and influential Joseph Banks, a Commissioner who had been a friend and supporter to Arnold. In 1803 Earnshaw produced a printed description of his time-keepers at the Board's request, as did John Roger Arnold, the son of his late rival. The first edition of this description contained some highly inflammatory remarks, mainly directed at Arnold, which were so inappropriate that almost all of the copies were recalled and destroyed.

An outraged Joseph Banks published a 'Protest against a vote of the Board of Longitude granting Mr Earnshaw a reward for the merit of his time-keepers' in 1804. Had Earnshaw attempted to challenge the formidable Banks he would probably have destroyed himself. Fortunately for Earnshaw, he had at least one supporter and that was Nevil Maskelyne, who took up the cudgels on his behalf and published a response to Banks's pamphlet.

Despite the public nature of Earnshaw's campaign to defend his claim of invention and the awkwardness that this must have brought between Maskelyne and Banks, the Astronomer Royal had great faith in Earnshaw's abilities and – if Earnshaw's writing is to be believed – one gets the impression that he found the hot-headed horologist entertaining company.

As with Arnold's and Coombe's before, Earnshaw's chronometers performed well at the Observatory and it was not long before he, too, was offered work. In 1792 Earnshaw cleaned and repaired K1 for the Board of Longitude and Maskelyne had another task in mind for him.

Maskelyne was again acting as adviser, looking for suitable instruments for the Armagh Observatory on behalf of its founder, Richard Robinson, Archbishop of Armagh. George Margetts, chronometer maker and mathematician, had already made a regulator for Armagh at Maskelyne's request but it failed to meet the required standard. Earnshaw described a conversation with Maskelyne about the Armagh commission. We get a good sense of his style of narration when he comments, without explanation, that he had been 'pirated' by Margetts and continued: 'The Dr. replied, that he had never met with the like; I was rival to every body!'[21] From Earnshaw's account, Maskelyne persuaded him to try his hand at making a regulator for Armagh by flattering him that his watches went so much better than those of his contemporaries. Earnshaw tells us that he at first resisted, saying 'I had never had made a clock, and did not know how many wheels were in one.'[22]

Maskelyne's faith in Earnshaw was not misplaced. The clock is of the same basic design formulated by George Graham earlier in the century and it is arguably the finest regulator of this type ever to have been made. The testimonials from the Reverend Dr James Hamilton are quite extraordinary, telling of a six-month deviation of only two-tenths of a second in that time. Earnshaw went on to make around a dozen regulators, one of which can be seen at the Royal Observatory today. The dial has been re-signed 'Redfearn' at some point in its history but the clock conforms closely to the description in the sale catalogue for Earnshaw's shop regulator, when the contents of his shop at 119 High Holborn (its site now commemorated with a 'blue plaque') were sold in the mid-1800s. In the same manner as the transit clock at Armagh, the clock has a near-hermetically sealed case with heavy brass hand screws to close up the joints, minimizing the effects of movement of air and ingress of dust and other evils, such as spiders.

An exciting recent acquisition (2009) to the National Maritime Museum's collection is Nevil Maskelyne's own pocket chronometer, Earnshaw No. 309, which is an interesting watch in its own right as it highlights a peculiar battle fought by its maker. In order to render the case as effective as possible in keeping the movement free of dust (and perhaps the curious owner's investigations), the dial and movement are secured by three tiny screws. When the watch is removed from its case, it is unusually plain in its appearance. It was purposely left so to counter claims that Earnshaw's watches only went well because of the extreme effort that went into finishing them, rendering the design less than practical. Earnshaw was not one to back down from an argument and so produced a series of watches with a minimal amount of finishing in order to silence his antagonists. The watches did perform just as well but, of course, he was then open to easy criticism that they were not very nicely finished (Fig. 5).

Earnshaw No. 309 appears at the top of an inventory of watches in the Maskelyne household, dated 27 April 1807. Of the eight listed, two are made by Earnshaw and another two were purchased from him a year earlier. One other watch on the list by an unspecified maker was on loan to 'Barton' (probably the computer Charles Barton). It seems from this inventory that Maskelyne had considerable faith in Earnshaw and that the mercurial watchmaker's lengthy protestation and supplication had not sullied their good relationship.

Fig. 5: The movement of Maskelyne's pocket chronometer by Earnshaw, with an intentionally plain finish.

Based on the success of the Armagh clock, it is perhaps surprising that Earnshaw was not commissioned to make an astronomical regulator to replace Graham 3. He addressed this in his appeal:

> After I had pointed out to Dr Maskelyne the absurd manner in which the famous Mr Arnold had jewelled the transit clock at Greenwich, he wanted me to re-jewel it, and do any thing else I thought necessary; I refused, saying, that let me do what I might, it was still Graham's clock, as his name was on it, and if the Royal Society, after the proofs I had given of my superiority, did not choose to order as good a clock of me, as I could make, they might keep their old one as it was, a standing monument of disgrace to Mr Arnold and others, who had botched it up in the manner it now is.

Within this particularly eloquent but undeserved poke at Arnold, it seems as though the truth was that the qualitative advantage in a new clock still did not outweigh the cost of a new one. Earnshaw did make adjustments to Graham 3 as well as the other two Graham clocks at Greenwich, by reducing the load of the motion-work of all three (Fig. 6).

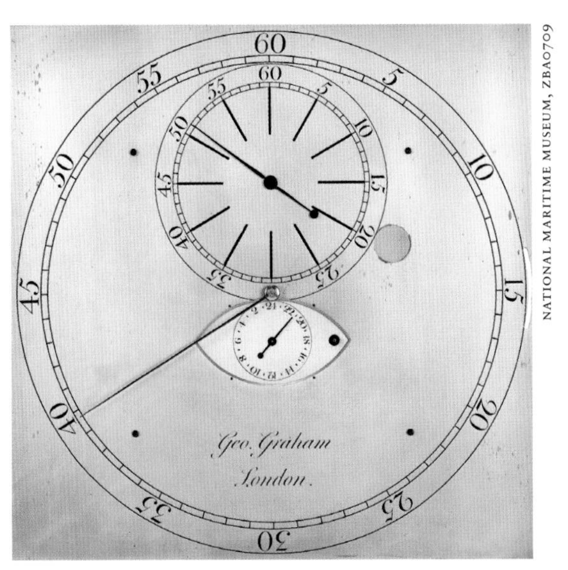

Fig. 6: Detail of Graham 3. The sharp elliptical aperture in which the hour hand runs is a marker of Earnshaw's simplification of the motion work and features on the three Graham clocks at Greenwich. By reducing one wheel in the motion work, friction is reduced and as a consequence the hour hand rotates in a counter-clockwise direction.

WILLIAM HARDY

Maskelyne remained satisfied with the Graham 3 until 1807, when William Hardy (d. 1832) was commissioned to make a new Observatory clock. He had already established himself as a talented and inventive horologist by gaining prizes from the Society for the Encouragement of Arts, Manufactures and Commerce for improvements to marine chronometers and devising a new type of clock escapement. Hardy's is known as a constant-force escapement, the principal advantage of this being that any change in power transferred through the clock's wheel train has a much reduced effect on the timekeeping.

Hardy wrote to the Secretary of the Board of Longitude in March 1807 about his invention and respectfully requested that it consider a proper trial of the escapement at the Royal Observatory. Feasibly prompted by John Pond's analysis of instrumental error at the Observatory in 1806, Maskelyne, on learning of the new escapement, paid a personal visit to Hardy and, by mid-May, had moved the latter's new clock to Greenwich. The trial began in late May when the clock's rate was checked by observation and the transit clock, alongside thermometer and barometer readings, and proved impressively stable. Maskelyne was so impressed that he persuaded the Board of Ordnance that Hardy was the best candidate to make a new clock to accompany the six-foot mural circle which was being built for the Observatory by Edward Troughton. He had high hopes for the new instruments and considered that they might supersede the old transit instrument and clock.

Hardy started making the new clock with great enthusiasm, sparing no expense or effort. He made the cutters that would produce geometrically correct epicycloidal tooth forms and sent them to Leyland of Prescot, Lancashire, who constructed the raw movements.[23] Hardy's work is astonishing: even when the wheels are set in free motion today the transition of power is incredibly smooth and almost silent. Charles Frodsham, a preeminent horologist, commented: 'The wheelwork in Hardy's regulators is amongst the best in England and the shape of the wheels and pinions makes the most perfect gearing I have ever witnessed.'[24]

Hardy tells us that Maskelyne treated him kindly and hospitably, encouraging him to make the best clock possible, that 'the order for the clock was unlimited in price, and [that he] directed me to consult [him] on every particular connected with it'. Sadly, Maskelyne did not live to see the clock installed in the Royal Observatory and this caused Hardy considerable trouble when it came to settling the bill of £325. Pond, as Maskelyne's successor, was not impressed: he referred it to the Royal Society and, after a lengthy process, the payment was reduced to £200. The London watchmaker, Alexander Cumming, was asked to act as a mediator and in correspondence on the matter expressed the view that making a clock for the Observatory was an honour and, as a prestigious contract, should not be about financial remuneration. Cumming's advice proved sound as Maskelyne had recommended Hardy's design to the Swiss surveyor, Ferdinand Rudolph Hassler, who in 1812 ordered two of his clocks for the American coastal survey.

Hardy found business at the Royal Observatory considerably different under Pond as the new Astronomer Royal. Following a visit in 1818 to set the clock Maskelyne had ordered in the Observatory, Hardy wrote a letter to the Secretary of the Board of Longitude, objecting to the 'careless indifference' he had met with there.

Hardy's design was excellent but did have a serious Achilles' heel in that its good performance was entirely dependent on the pallets being lubricated with fresh, clean oil. It therefore became troublesome when this was not provided at the Observatory, and the final insult to Hardy came when the ambitious clockmaker Edward John Dent (1790–1853) was asked to deal with the matter. Dent removed Hardy's escapement and the majority of the train jewels and, as if that was not enough, had his own name engraved across the dial (Fig. 7).

Fig. 7: The dial of William Hardy's regulator bearing Dent's inscription.

The negative image of Maskelyne stems exclusively from interaction with the Harrisons and Thomas Mudge junior and it seems likely that he initially, like many of his contemporaries, viewed the marine timekeeper as secondary to the lunar-distance method of finding longitude. Maskelyne, in his account of the observatory trial of Harrison's watch, conceded that while it

> could not be depended upon to keep the longitude to within half a degree in a West India voyage of six weeks ... [it was] nevertheless, a useful and valuable invention, and in conjunction with the observations of the distance of the moon from the sun and fixt stars may be of considerable advantage to navigation.[25]

The main complaints against Maskelyne stem from poor handling or misunderstanding the limits of the timekeepers. They came from men who knew the intricacies and weak points of their machines and viewed Maskelyne as unfit to handle them. He was perfectly capable of setting up a portable astronomical clock and so one would think he must have had a fair practical understanding of clockwork, but regular reports in his memoranda on the performance of William Coombe's clock cast some doubt on this.

Maskelyne notes, in August 1792, that Earnshaw set the case upright and put the clock in beat.[26] It is very surprising that Earnshaw should have needed to do this, for any precision clock should be fixed rigidly to the wall if it is expected to keep time reliably. The Coombe clock continued to perform erratically and it was not until 12 September 1798 that it was eventually set upright again and fastened to the wall to prevent its pendulum from hitting the side of the case.[27]

Unlike the pendulum clock, marine timekeepers are portable and Maskelyne probably expected them to work well on demand, regardless of condition or any motion to which they might be subjected. This naïve expectation is best demonstrated in January 1772, when Kendall's K1 copy of H4 stopped and Maskelyne describes how he 'could not make it go again ... tho' I warmed it by the fire and gave motion to the balance', treating a sophisticated machine that is compensated for changes in temperature almost as if it were an infant.

While clearly a practical man, it seems that Maskelyne was not mechanically minded and this view is reinforced by the fact

Fig. 8: 'K1' by Larcum Kendall, 1769.

that his only 'hands-on' contribution to horology was mathematical calculation at the request of Arnold.[28] He naturally desired more accurate clocks to work with and, from his memoranda, we learn that he took great pleasure from the good performance of watches owned by his family, regularly jotting down their rates measured against Coombe's clock.

Despite any shortcomings when it came to handling clocks, it is fair to say that his encouragement and guidance catalysed some of the finest precision clocks of the eighteenth and early nineteenth centuries and this contribution to horology absolutely outweighs the negatives. As the last clockmaker to have benefited from Maskelyne's enthusiasm for improvement, it seems appropriate to let William Hardy have the final word on Maskelyne:

> such was his perseverance, that he was out there in the most inclement of weather, superintending the workmen, which I believe tended much to shorten his days: he was most indefatigable in his application to the interest of the Observatory, no doubt, for the great end of advancing the science of astronomy, in which he delighted.[29]

FURTHER READING

There is no better overview of the astronomical clocks and watches used in the transit of Venus expeditions of the 1760s than in Derek Howse and Beresford Hutchinson's *The Clocks and Watches of Captain James Cook, 1769–1969* (Antiquarian Horological Society: reprinted from the quarterly issues of *Antiquarian Horology*, 1969). For further technical and biographical detail regarding astronomical clocks and their makers, Derek Roberts' *English Precision Pendulum Clocks* (Pennsylvania: Schiffer, 2003) discusses some of the work of makers mentioned in this chapter and illustrates Maskelyne's clock by William Coombe.

Vaudrey Mercer's book *John Arnold and Son, chronometer makers, 1762–1843* (London: Antiquarian Horological Society,

1972) gives a good biographical account of the maker and his work. For an extensive catalogue of Arnold's clocks and watches, Hans Staeger's volume, *100 Jahre präzisionsuhren von John Arnold bis Arnold und Frodsham 1763–1862* [100 years of precision timekeepers from John Arnold to Arnold and Frodsham] (Filderstadt: H. Staeger, 1997) is well worth seeking out and the pamphlets discussed in this chapter are reproduced in full there. Arnold's vociferous opponent, Thomas Earnshaw, provides some extraordinary background detail in *Longitude. An Appeal to the Public.* This is a rare book but the British Horological Institute issued a reprint in 1986.

Excepting William Hardy and Coombe, all of the makers discussed here feature in William Andrewes, ed., *The Quest for Longitude* (Cambridge, Mass.: Harvard University Press, 1996) among a wealth of other information relating to the longitude story.

CASE STUDY E

INSTRUMENTS OF EXPLORATION

As a combined result of three circumstances, Maskelyne was to become a central figure in the planning and organization of voyages of scientific exploration. First was his personal experience of maritime travel and expeditionary astronomy. Secondly, he had shown himself at the Royal Society to be capable of helping to equip and issue instructions for such work, on the occasion of the 1761 transit of Venus and subsequently with the Society's involvement in the fixing of the boundary between the colonies of Pennsylvania and Maryland – the Mason-Dixon line, named after the same Charles Mason and Jeremiah Dixon who had observed the transit from the Cape of Good Hope. Finally, as Astronomer Royal and Commissioner of Longitude, Maskelyne was an obvious figure for the Admiralty to consult and the chief authority within the Board regarding its growing stock of hardware.

It is notable that, in his involvement in many of these expeditions, Maskelyne was adept at taking the opportunity to push a number of linked projects, just as he had done personally on the St Helena voyage. He had also leapt on Mason and Dixon's suggestion that their land survey should be turned to scientific account by going on to measure a degree of latitude, thus contributing to the project to determine the figure of the Earth. Maskelyne's openness to combined objectives related to his range of interests, but it was also very clearly the result of the inspiration provided by the published accounts of the expeditions of Picard, Cassini, Bouguer, La Condamine, Maupertuis, Boscovich and Lacaille over the previous half century. It was, in addition, the best way to take advantage of the limited funds available for scientific work: exploration was the channel through which science and empire-building helped legitimize each other. Thus astronomy could be combined with navigation, geodesy, gravimetric, hydrographic and geomagnetic observations, as well as the testing of novel instruments and techniques.

It was through the Royal Society, and as Astronomer Royal, that Maskelyne became involved with the first of Lieutenant James Cook's voyages of exploration, in the *Endeavour.* This was because it was, first and foremost, one of the five British expeditions dispatched to observe the 1769 transit of Venus. Experienced from the previous transit and now Astronomer Royal, Maskelyne was the key figure within the Royal Society's planning committee, formed to choose locations, equip the expeditions, select observers and give instructions for setting up temporary observatories in order to check the going of clocks, fix positions and observe the transit itself. Most of the observers were known to Maskelyne, either as assistants at the Royal Observatory or from having worked as computers of the *Nautical Almanac.* Cook was the apparent outsider but he had already proved himself a highly competent surveyor and had his observation of a solar eclipse published in the Royal Society's *Philosophical Transactions* in 1767. His observations had been compared to others by Thomas Hornsby and communicated to the Society by John Bevis, both of whom were very familiar to Maskelyne, the former as a Commissioner of Longitude and the latter as an old friend and fellow member of the Transit of Venus Committee.

It is well known that Cook's voyage also had instructions to go beyond the observing location, Tahiti, and seek out a supposed great southern continent. The most high-profile scientific activity of the expedition, after the transit observations, was the botanical, ethnographical and zoological collecting led by Joseph Banks. However, the Royal Society-funded astronomer on board, Charles Green, who had accompanied Maskelyne to Barbados, was also to play a significant role in aiding navigation, in fixing positions on land astronomically to enhance the accuracy of survey work and, as and when opportunity arose, in a whole range of other observations. Maskelyne made sure that the voyage was supplied with Tobias Mayer's lunar tables and the newly available copies of the *Nautical Almanac* for 1768 and 1769. Cook was to give lunar-distances a good report, saying that it was 'a Method which we have generally found may be depended on within 1/2 degree', and that it was greatly assisted by the *Almanac*, which would be even more beneficial if it were published further in advance. He noted, too, that Green had taken care to teach others the method and that 'by his instructions several of the petty Officers can make and calculate these observations almost as well as himself'. Cook added that 'It is only by such Means that this method of finding the Longitude at Sea can be put into universal practice.'[1]

The other expeditions for 1769 were to Hudson's Bay, to which William Wales and Joseph Dymond were dispatched, to Norway (William Bayly and Jeremiah Dixon) and, rather closer to home, Charles Mason to County Donegal and John Bradley (James Bradley's nephew) to Cornwall. Each was supplied with the same main set of instruments, made by identified and admired instrument makers. The two-foot reflecting telescopes should be by James Short or John Bird, the twelve-inch astronomical quadrants by Bird, and the clocks by John Shelton or John Ellicott. Maskelyne wrote the formal instructions for each observer and also added a long appendix to the 1769 *Nautical Almanac* to aid any other independent observer who happened to be in the right place at the right time on 3 June 1769.

The 1769 expeditions reused the instruments that had been bought for 1761, as well as requiring more to be bought. This increased the stock of instruments belonging to the Royal Society that were available for future use, just as the Board of Longitude was beginning to make its own collection. The availability of hardware was behind Maskelyne's next, significant step. When the Admiralty planned another voyage for Cook, Maskelyne took the opportunity to write to Lord Sandwich, who was First Lord of the Admiralty (and therefore, *ex officio* a Commissioner of Longitude) on 25 October 1771. He suggested that the planned voyage to the South Seas:

> May be rendered more serviceable to the improvement of Geography & Navigation than it can otherwise be if the ship is furnished with Astronomical Instruments as this Board hath the disposal of or can obtain the use of from the Royal Society and also some of the Longitude Watches; and, above all, if a proper person could be sent out to make use of those Instruments & teach the Officers on board the ship the method of finding the Longitude.[2]

This suggestion was approved by the Board and taken up by the Admiralty. Thus Larcum Kendall's first watch, K1, and three timekeepers by John Arnold went with William Wales and William Bayly, together with a range of other instruments belonging to the Royal Society and the Board, for use at sea and on land. If the Harrisons had been willing, H5 would have gone too. Again, the choice of observer, provision of instructions and instruments were down to Maskelyne. He was also the person to whom the observations made were sent, the results ultimately being published by the Board.

Thus the pattern was established. The sending of observers and instruments, including marine timekeepers, on Cook's third voyage (1776–80) was again initiated by Maskelyne writing to Sandwich. When the idea was endorsed by the Board, he was then encouraged to list the instruments that he 'shall judge proper and necessary'.[3] On this occasion, Cook and William Bayly were already waiting outside the meeting room to hear about the decision and the salaries offered. Maskelyne's draft list of instruments and books for this voyage – Cook's last – survives in the library of the National Maritime Museum (Fig. 1).[4]

The instructions, as before, included making daily on-board observations for latitude and longitude (by lunar distance and watch), of compass variation and magnetic dip, of air temperature and of the salinity and temperature of the sea. Bayly was also to make remarks on the existence or nature of the Southern Lights, to teach those officers who wished to learn the use of the instruments and the lunar-distance method, and to establish the

List of Instruments be ~~belonging to the Board of Longitude~~ delivered to Capt. Cook May 27 1776

X An Astronomical Clock
X An alarum Clock
X An astronomical quadrant
A Hadley's Sextant by Ramsden
A Hadley's Sextant by Dollond
A transit instrument of 4 feet, with its level & stand, to Mr. Bayly only.
X An achromatic telescope with a treble object glass of 46 inches focus, with an ~~divided~~ object glass micrometer, and an eye tube with moveable wires this latter to Capt. Cook's only.
X A Reflecting telescope
X A 4 feet hand perspective with a large aperture.
X A marine dipping needle, with 6 magnetic steel bars
Two small variation compasses
X An azimuth Compass with a spare card.
X A theodolite & Gunter's chain, to Capt. Cook only
A basan for holding quicksilver for observing double altitudes with 6 lb of quicksilver.
Two portable barometers
X Six thermometers
Kendal's first watch made in exact imitation of Mr. Harrison's to Capt Cook, the 3d made do to Mr. Bayly
A pinchbeck pocket watch with a second hand and ruby cylinder, to Capt. Cook only.
X A Marine barometer
X A ~~wooden~~ bucket contrived for fetching up sea water from great depths for trying its saltness as well as its degree of coldness, with 2 thermometers belonging to it
3 bottles for weighing salt water in
X A hydrostatic ballance.
X Two night telescopes. X Tent Observatory

List of Books.

3 Books of Folio Tables of Refraction & parallax
x 2 Mayers Tables
x 2 Hazleden's Seaman's daily Assistant
1 nautical almanac of 1769, 1 of 1772 & 1 of 1773
x 6 nautical almanacs of 1776, 1777, & 1778
2 requisite Tables to Mr Bayly only
3 Gardiner's Logarithms, ... published at Avignon. 1 to Capt Cook, 2 to Mr. Bayley

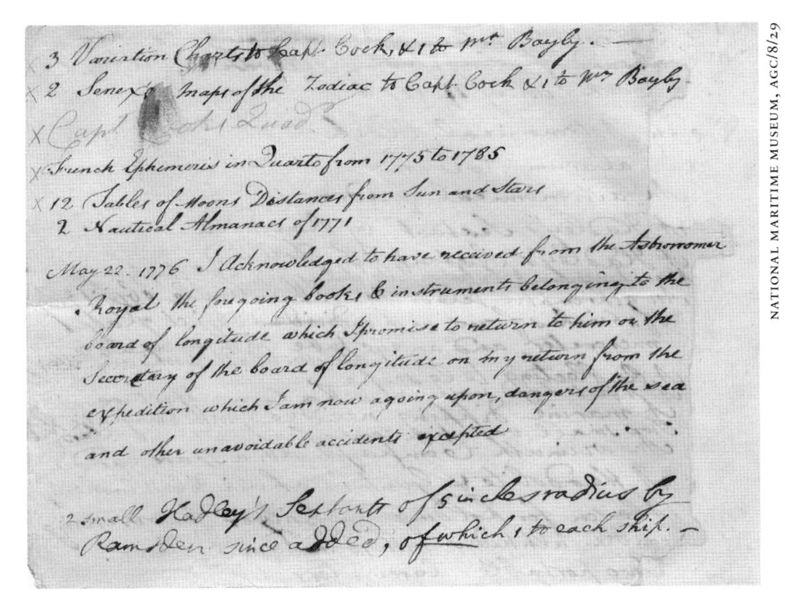

3 Variation Charts to Capt. Cook, & 1 to Mr. Bayly. —
2 Senex's Maps of the Zodiac to Capt. Cook & 1 to Mr Bayly
~~Capt Cook 1 Quad~~
French Ephemeris in Quarto from 1775 to 1785
12 Tables of Moons Distances from Sun and Stars
2 Nautical Almanacs of 1771

May 22. 1776 I Acknowledged to have received from the Astronomer Royal the foregoing books & instruments belonging to the board of longitude which I promise to return to him or the Secretary of the board of longitude on my return from the expedition which I am now agoing upon, dangers of the sea and other unavoidable accidents excepted

2 small Hadley's Sextants of 5 inches radius by Ramsden since added, of which 1 to each ship. —

Fig. 1: A draft list of instruments for Captain James Cook's third and final voyage, drawn up by Maskelyne on 22 May 1776. This shows the extent of specialist navigational, surveying and scientific equipment and books taken on such voyages, as well as Maskelyne's ongoing role in relation to them.

position of headlands, island and harbours. When on land there was the same routine, plus observations of the clock's pendulum with reference to local gravitation. There were also detailed instructions on how the observers should keep their records.

As Joseph Banks became more powerful in the Royal Society, becoming President in 1778, it is clear that he was able to take a more direct role in pressing for particular voyages to take place. Nevertheless, Maskelyne remained largely in control of selecting astronomers and the hardware of physical observation. Thus his instruments and instructions went out with Israel Lyons on the north-bound voyage of Constantine John Phipps (1773), with William Dawes on the First Fleet to Australia (1787), with William Gooch when he was sent to join George Vancouver on the north-west Pacific coast (1791), and with John Crosley, sent to join Matthew Flinders' circumnavigation of Australia (1801). It is clear, though, that the methods indicated by Maskelyne, and taught at sea by Green, Wales and others, also began to filter into at least specialized naval practice. Individuals like William Bligh, Vancouver and Flinders learned much while at sea on Cook's voyages, and they were themselves entrusted with Board of Longitude instruments, including the precious early timekeepers.

Novice observers, like the young Gooch, had, however, to be vetted and trained by Maskelyne. Thus, before he could be formally proposed as a possible Board of Longitude observer, as had been suggested by the Cambridge mathematician, Samuel Vince, later Plumian Professor of Astronomy, Gooch went to Greenwich. On 20 April 1791 he wrote to tell his parents 'I find Dr. M. a very pleasant man; he doesn't seem to intend to examine me, but to be perfectly satisfied with Vinces word'.[5] Yet Gooch did note that Maskelyne 'attended closely to every thing I undertook for Practice; and observ'd the accuracy of my observations by seeing what they were & calculating what they should be & then seeing how near they agreed'.[6] This work with the Observatory's astronomical regulator and transit telescope was enough to convince Maskelyne of Gooch's suitability, despite opposition from Banks: on 11 June, Gooch was at the Admiralty, sitting outside the Board of Longitude meeting at which he would be appointed.

Gooch's letters also reveal how he was advised and prepared by Maskelyne and the experienced Wales. Some of this assistance was practical and financial, with Wales making suggestions about how to make money out of trading fur during his travels and what furniture and clothing to purchase before going. He was also closely informed about his duties and responsibilities, both as the Board's astronomer and as the temporary guardian of its expensive suite of instruments. Maskelyne's precautions – the careful selection and training, the long instructions on observation and record-keeping, the use of instruments that were as nearly identical as possible, made by the best makers – were all intended to overcome the problems of trusting the value of observations made by a range of individuals with various instruments at a great distance from home. However, Gooch's letters make clear that, despite this effort, observations with fragile instruments, made by fallible individuals in the difficult conditions of sea travel and distant lands, could never be completely satisfactory

Sadly, Gooch was killed in Hawaii before he ever reached Vancouver's expedition. Vancouver and his crew were able to make use of the Board's instruments, but it was up to Maskelyne to do his best to recover Gooch's belongings and money owed to him. The archive reveals that, as so often, Maskelyne went above and beyond the call of duty in dealing with those for whom he had become responsible. He undertook extensive correspondence around London and overseas, tracking down what was owed to Gooch, detailed in letters to the dead man's father. He went so far, for example, as to carry out lengthy investigations into the law and custom surrounding the purser's subtraction of a hefty percentage for dealing with the sale of some of Gooch's belongings. When Gooch senior enquired about Maskelyne's expenses, he was quick to respond that he had incurred few and wanted nothing for his time.[7]

The archives and published observations from such voyages bear testament to Maskelyne's active involvement in this period of scientific exploration. It helped underpin an alliance between naval surveying activity and the metropolitan scientific world, and accustomed certain officers and parts of the navy to the use of precision instrumentation. These things were to continue in the following century and, while the voyages of Charles Darwin, Joseph Hooker and Thomas Huxley can be seen as having been inspired and informed by that of Joseph Banks on the *Endeavour*, we can trace Maskelyne's and Cook's influence in the range of work that was carried out by surveyors and men of science like Francis Beaufort, Robert FitzRoy and John Lort Stokes.

Like so many others involved with the exploration and surveying of coastlines in the eighteenth and nineteenth centuries, Maskelyne is also recalled by a number of place names given by the astronomers whom he had selected and the surveyors he had equipped. There is a Maskelyne Point and Maskelyne Island in British Columbia, Canada, named by George Vancouver in the 1790s. The peninsula (originally known as Tar-ra and Tullagalla) near Sydney, New South Wales, where William Dawes established his observatory after the arrival of the First Fleet in 1788, was named Point Maskelyne – although after Dawes' departure it was renamed Point Dawes. The Maskelyne Islands of Vanuatu in the South Pacific were named by William Wales during Cook's second voyage. While the giving of such names was often just for form's sake, Wales very touchingly wrote that he had

> ventured to call them by the name of a person to whom I owe very much indeed; one who took me by the hand when I was friendless, and never forsook me when I had occasion for his help; and who, I hope, will not be offended at this public acknowledgement of his favours.[8]

6

'HUMBLE SERVANTS', 'LOVING FRIENDS', AND NEVIL MASKELYNE'S INVENTION OF THE BOARD OF LONGITUDE

Alexi Baker

When Maskelyne was appointed Astronomer Royal in 1765, he became a pre-eminent European authority in the field that had been his passion since childhood, and master of all of the activities and international networks of communication centred upon Greenwich. He also automatically became a Commissioner of Longitude. Astronomy was vital to early modern navigation and the Royal Observatory was specifically founded with navigational advancement – and in particular the determination of longitude – in view. As a result, Maskelyne followed in the footsteps of well-known predecessors like John Flamsteed and Edmond Halley in serving as a national and international authority on longitude while at Greenwich.

He diverged dramatically from them, however, when it came to the way in which he conducted himself as a Commissioner, doing more than any other single official to forge a standing Board of Longitude from the largely independent Commissioners who existed after the Act of 1714. Maskelyne also helped to make the Board a prominent national and international force for improving and carrying out navigation, exploration, science and technology. This institutionalization and expansion were largely fuelled and shaped by his wide-ranging networks of personal and institutional ties. These were embodied in the endless letters that he exchanged with men, and occasionally women, who often styled themselves his 'humble servants' or 'loving friends'. He also established a more regular system than his predecessors for responding to unsolicited longitude-related correspondence. For all of these reasons it was fitting that, as he later recalled, Maskelyne's first meeting as a Commissioner of Longitude coincided with his first day as Astronomer Royal.

Today it seems most natural to think of position-finding in terms of technology, whether it be GPS (Global Positioning System) or marine chronometers. However, for centuries the improvement of navigation, and thus of estimating longitude at sea, was considered an astronomical problem. New technologies were intended to make, or to be used in conjunction with, astronomical observations and publications. This essentially remained the case until the spread of GPS at the end of the twentieth century. Before that, sextants were still in use even on high-tech warships to observe the positions of the Sun and stars.

This resulted in the Astronomer Royal being considered one of the premier longitude experts in England, and in practice the leading one, from the creation of the post in 1675. The Astronomers Royal were often consulted by individuals and by governmental, commercial and intellectual institutions about issues and ideas related to longitude and to general navigation. Sometimes they examined related inventions and proposals, or attended trials of them on the Thames and at sea. They also made astronomical observations themselves, with varying degrees of regularity and openness, which might improve techniques and predictive tables.

Sometimes the Astronomers Royal made observations far afield as well in pursuit of the longitude before they took up the position at Greenwich, as would Nevil Maskelyne. (Early modern astronomers were often required to be far more adventurous than their modern counterparts, sometimes trekking to distant viewing points and even going behind enemy lines during war!) At the turn of the eighteenth century Edmond Halley twice commanded the *Paramour* – a small three-master that was the first purpose-built naval research ship – and carried out investigations in the Atlantic into the differences between geographic north and magnetic (compass) north. He endured trials including crew insubordination and being mistakenly fired on as a pirate. His efforts produced a ground-breaking magnetic-variation chart in the hope, periodically entertained throughout the 1600s and 1700s, that the variation was predictable enough to aid in estimating position at sea (Fig. 1).

Halley would later turn to lunar methods instead, but these efforts no doubt contributed to his appointment as the second Astronomer Royal from 1720 to 1742 – just as Maskelyne's efforts at trialling and improving the lunar-distance method would put him in good stead in the 1760s.

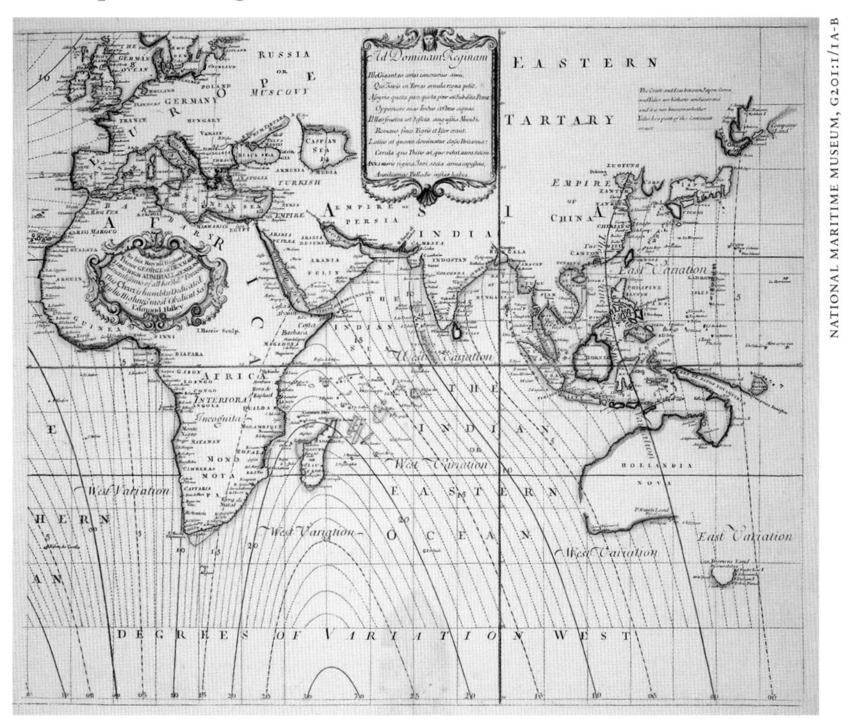

Fig. 1: 'A new and correct sea chart of the whole world shewing the variations of the compass' by Edmond Halley, 1702.

THE COMMISSIONERS OF LONGITUDE

The developments of the sixteenth and seventeenth centuries, including Halley's work on magnetic variation, helped to set the scene for legislators to listen when William Whiston and Humphry Ditton began to lobby for a public longitude reward in 1713.[1] The Act of Parliament passed as a result in 1714, appointed twenty-four Commissioners either by name or office, and later Longitude Acts would increase this number as the original named officials passed away. These mainly included key figures from politics, the Royal Navy, astronomy and mathematics – groups which were already perceived as being some of the key practitioners and beneficiaries of the search for longitude. Naturally, they included the Astronomer Royal. The details of the 1714 Act were largely influenced by the suggestions of the men who had recently testified to Parliament, including Isaac Newton and Edmond Halley, and by two English precedents. These were Charles II's appointment of Commissioners to examine a longitude proposal in 1674, which contributed to the foundation of the Royal Observatory, and a reward established by bequest when Thomas Axe, a Somerset gentleman, died in 1691.

However, the legislation did not, itself, set up a proper standing body or 'Board of Longitude' to deal with the multitude of proposals that were likely to be made in response to the rewards. The legislation only named individuals as acceptable judges, or Commissioners. Before the modern era, that title could signify an individual who was granted a responsibility or commission. For example, a representative of the English King for various purposes was often labelled a King's Commissioner. The individual Commissioners of Longitude were given limited guidance, no official communal meeting place or existing funds, and no salaries to make up for time lost from their 'day jobs'.

It appears that many members of the public did not expect there to be a standing longitude body either: for example, such an institution was not referred to in the limited coverage of the first-known communal meeting of Commissioners on 30 June 1737. The *London Evening Post* simply reported it as a meeting of relevant individuals: 'the Right Hon. Arthur Onslow, Esq; Speaker of the House of Commons, the Right Hon. the Lords Monson and Lovel, [Admiral] Sir John Norris, [Admiral] Sir Charles Wager, and several Persons of Distinction, view'd a curious Instrument for finding out the Longitude, made by Mr. [John] Harrison'.[2] (The instrument was Harrison's 'H1' timekeeper, now at the National Maritime Museum.) The earliest example found to date of a clear use of 'the Board of Longitude' in the sense of a standing body, rather than of the attendees at a specific meeting, is in a letter from the Astronomer Royal, James Bradley, in 1756. This was more than forty years after the establishment of the longitude rewards.

Comments made by Commissioners, reward-seekers, and interested parties such as Newton suggest that it had been up to the individual post-holders to determine the direction of their service after 1714. We can also see this in how, decades later, Nevil Maskelyne was able to play such a key role in redefining the Board of Longitude and its activities by bringing to bear his personal interests, dedication and energy. It does not appear that any of the founding officials pushed for communal action or dramatically altered how they had always responded to proposals. This may have in part been because the death of Queen Anne soon after the passage of the Act altered the political landscape and disempowered some of the individuals named.

There was clearly public confusion, which persisted for decades, over how best to put forward a proposal for the rewards. Specific examples of this appear in letters to and from related individuals including the Astronomers Royal, queries and postings in newspapers and periodicals, and sometimes reward-seekers' pamphlets and books. A very interesting example is that of Jane Squire (*c*. 1686–1743), the only woman to have openly pursued the longitude rewards.[3] Squire tried over at least twelve years to receive a judgement on her religiously motivated scheme from the Commissioners, and published two books in 1742 and 1743. The books included letters that she reported exchanging with some of those officials and with other European intellectuals.

Squire had the Attorney General read her the Act of 1714 and inferred from it that she should present her ideas to one of the current post-holders. She first approached the President of the Royal Society, Hans Sloane, who kindly suggested that she consult Halley, by then Astronomer Royal. He did not respond to her directly, so she proceeded to send a printed copy of her proposal to each Commissioner's home in August 1731 – again receiving no response.[4] In early 1733, Squire exchanged letters with her acquaintance Sir Thomas Hanmer, Bt., who was one of the few original named Commissioners still living. Hanmer suggested that she publish her ideas because he could not see his colleagues judging all first-round proposals, as she had expected. Instead, an idea 'must undergo the Scrutiny of all the great Professors of the Sciences of Astronomy and Navigation, and not only that but it must stand the Test of Practice'.[5]

As we can see through examples like this, the dynamics of the search for the longitude in Britain after 1714 were very similar to those before – but with a much greater level of public participation, at home and from abroad. In lieu of there being a standing body – or even one official mode of application for the new funds – reward-seekers contacted a range of likely-seeming experts, patrons and institutions: moreover, this dynamic would persist even after the development of a standing Board. Those contacted included intellectuals and professors, members of local and national government, the Admiralty and Royal Society, and occasionally institutions like the East India Company and other trading companies.

However, the Astronomer Royal remained the most approached and cited longitude expert – and, clearly, was often named as such by other Commissioners. It is most often he whom we find being sent, examining, and at times responding to proposals, and also approving inventions and techniques for sea trials. For example, the young self-taught peruke (wig) maker, John Bates, conducted a sea trial of his celestial method of finding longitude in 1728–9 with the reported approval of Halley and the Lords of the Admiralty. This was a rags-to-riches type of story which delighted the newspapers. Bates also approached the

East India Company in October 1730 to seek more funding, or perhaps a reward.[6]

Of course, the central importance of the Astronomer Royal to the search for longitude sometimes prompted accusations of conflict of interest, especially when the post-holder was working on a different method. This was perhaps inevitable, since the Crown – in creating the position – had directed its 'royal astronomers' to dedicate themselves to the improvement of finding longitude at sea. Jane Squire and other reward-seekers criticized Halley for pursuing a lunar method even as he acted as the ultimate judge of their proposals. On 24 December 1731 Squire complained to Admiral George Byng, Viscount Torrington, Commissioner by merit of being First Lord of the Admiralty, of a rumour that the astronomer would be given first chance at a reward owing to his years of effort:

> it would be the highest Injustice to all ... to determine them to be sacrificed, to the Honour and Interest of one ; who (by having, himself undertook to make this Discovery) is certainly disqualified to be the sole Judge of the Methods by which other propose it.... I know we boast the utmost Perfection in the first [astronomy], and have even idoliz'd a Name we think we owe it to ; but it still remains a strong Objection against our Astronomy, that it yet enables us not to find our Longitude.[7]

Similar suspicions were expressed against Nevil Maskelyne in the 1760s, and would be vociferously pursued by the Harrisons, as he made the transition from private advocate of the lunar-distance method to Astronomer Royal. This may be why a resolution appears to have been considered for the Longitude Act of 1765 that would have made all Commissioners ineligible to receive reward money (Fig. 2).[8] Maskelyne had, after all, become one of those officials while the legislation was still being thrashed out.

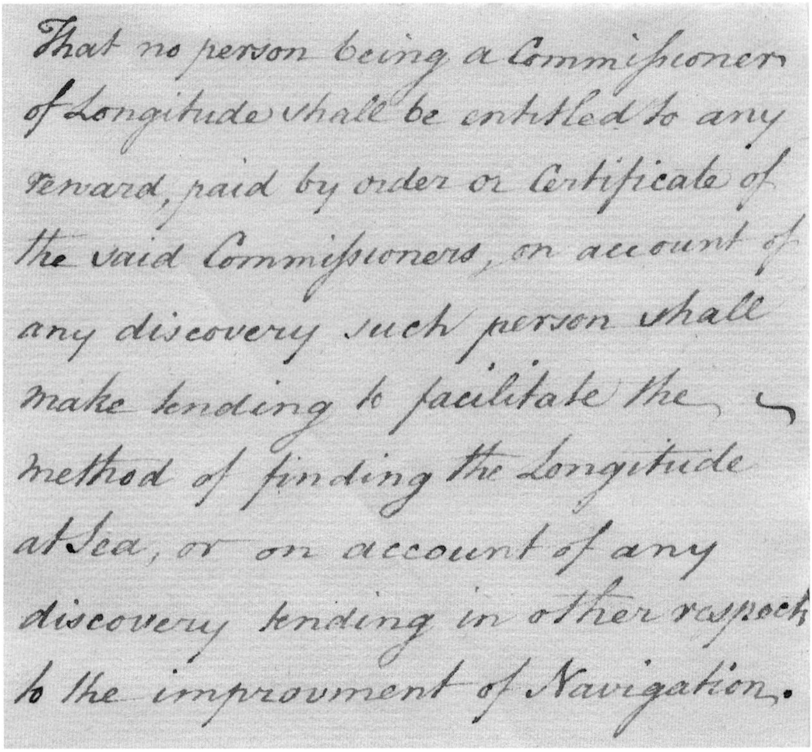

That no person being a Commissioner
of Longitude shall be entitled to any
reward, paid by order or Certificate of
the said Commissioners, on account of
any discovery such person shall
make tending to facilitate the
method of finding the Longitude
at Sea, or on account of any
discovery tending in other respect
to the improvment of Navigation.

Fig. 2: A note found among some documents belonging to Viscount Barrington, then Treasurer of the Navy, concerning Commissioners of Longitude being ineligible for rewards, c.1765.

THE BIRTH OF THE BOARD

On Maskelyne's accession as Astronomer Royal in 1765, he joined a Board of Longitude that had only recently become a recognizable standing body, and one increasingly known as 'the Board'. There is so far no evidence of communal meetings of the Commissioners before 1737. Eight are known to have taken place during the 1740s and 1750s, although each drew only a percentage of the named officials. It appears that it was the great public and institutional interest in the first marine timekeeper (1735) of clockmaker John Harrison that prompted this change in behaviour. The Commissioners worked closely and amenably with Harrison for at least the next twenty years, directing the bulk of their attention and expenditures to him.

However, early on they also began considering and supporting other proposals for improving longitude-finding and general navigation. These included William Whiston's magnetic variation observations, and Christopher Irwin's 'marine chair' to steady an astronomical observer at sea. During the 1750s, the Commissioners first began working with the lunar tables and technology devised by Tobias Mayer of Göttingen as well. These would greatly improve the lunar-distance method and result in the development of the sextant, both vital to Maskelyne's own advances with 'lunars'.

The 1760s were when the Commissioners truly began bureaucratizing and formed a standing Board of Longitude, no doubt in part because of the increasing general bureaucratization of government in Britain during that period. They began meeting at least annually and sought, for the first time, travel compensation for the professors and Astronomer Royal, and a small salary to hire a secretary for institutional records. (This is why the bulk of the surviving Board archives, which are now online, date from the 1760s or later.) At the outset of these developments, Maskelyne was not yet a Commissioner, although he had worked with them and with the Royal Society – frequent, and overlapping, institutional partners. Some of these activities were discussed in the same meeting, on 19 January 1765, at which the Commissioners heard he would shortly be joining their ranks as the new Astronomer Royal.

It is clear from a variety of records including correspondence, publications, and the surviving Board archives that, within only a few years, Maskelyne established himself as the most active Commissioner and as the nation's best-known longitude expert. This was no doubt in part because of the traditional association of longitude with the post of Astronomer Royal. However, he also took a different approach at Greenwich and as a Commissioner from those of his predecessors. Maskelyne was central to the reorganization of the Royal Observatory, as shown in Chapter 3 of this book. Previous astronomers had often refused to make the results of their observations public and sometimes let the buildings and instruments fall into disrepair. (Of Maskelyne's insistence upon publication, Delambre, Director of the Paris Observatory, said: 'for this he deserved to be the leader and, as it were, the regulator of astronomers for forty years.'[9]) For the first time, a number of the assistants who worked at the Observatory also went on to participate in geodetic and astronomical expeditions in their own right, owing to his teaching and influence. Maskelyne proved more dedicated to and organized in addressing longitude-related correspondence and activities as well. He also appears to have been the first post-holder interested in reshaping and strengthening the Commissioners.

Of course, there were other active Commissioners of Longitude during these decades too, such as Joseph Banks, who was President of the Royal Society from 1778 until his death in 1820. The Board might therefore have continued to expand and to bureaucratize without Maskelyne, but it seems unlikely that it would have achieved the same reach, or perhaps the same longevity. The astronomer initiated, organized and oversaw the body's activities to a remarkable degree during his forty-six-year tenure. This extended from the bureaucratic minutiae of its operations to the choice, planning and publication of its best-known international endeavours. The proposals and related correspondence received by the Board were most often turned over to him, too, for his expert opinion on whether or not they merited further attention.

Maskelyne was often more neutral and polite in his comments than were other eighteenth-century longitude experts, such as James Hodgson at the Royal Society and the earlier Astronomers Royal. In 1783, William Fuller of London sent the Board a handbill which described how he had published his method of finding the longitude by the fixed stars in the newspapers, and had taken copies to the Houses of Parliament, the Admiralty, and different coffee-houses. Although he was rather presumptuous in expecting the advertising to induce the Commissioners to contact him, Maskelyne did meet him – and then, as was common, noted succinctly on the back of the relevant documents that he was 'of opinion that they do not merit the attention of the Board of Longitude' (Fig. 3).[10]

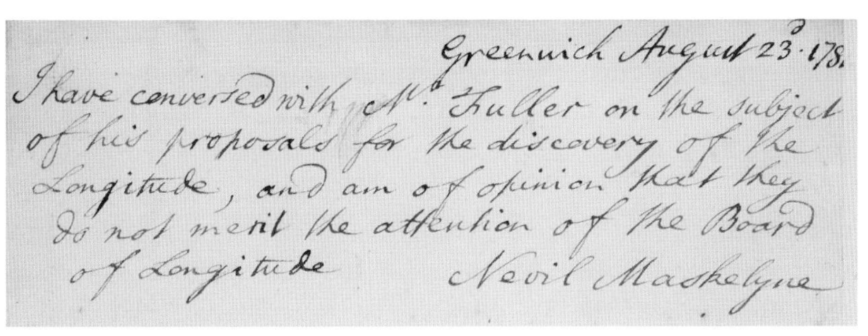

Greenwich August 23. 1783.
I have conversed with Mr. Fuller on the subject of his proposals for the discovery of the Longitude, and am of opinion that they do not merit the attention of the Board of Longitude
Nevil Maskelyne

Fig. 3: Nevil Maskelyne's annotation on the reverse of a letter from William Fuller to the Commissioners of Longitude on 6 March 1783, giving his opinion of the content.

In comparison, in 1733 James Hodgson scathingly commented upon the longitude proposal of clockmaker P.J. Rocquette, 'What answer must be given to a Man, who is so very ignorant of the first Principles of Astronomy and Philosophy, who has asserted so many falsehoods, and calls them Demonstrations; and is so vastly fond of his Performance.'[11]

Maskelyne also wrote notes and drafts for many of the Board of Longitude's minutes and for other documents, including background for its legal and legislative debates, produced a wide range of correspondence, and published volumes. Just a year after his appointment, as described in Chapter 4, he succeeded in initiating the production and publication of the *Nautical Almanac* under the auspices of the Board, and it has been published annually ever since. By 1768 he had Robert Bishop, a former Royal Naval master and pilot who repeatedly collaborated with the Board, produce printed forms to be used with the lunar-distance calculations. He further oversaw the production of a number of other publications for the Commissioners, including those related to Harrison's timekeepers and to the production by John Bird and Jesse Ramsden of precision 'scientific' instruments.

Maskelyne's activities and interests with the Commissioners, at the Royal Observatory, and with institutions such as the Royal Society constantly intermingled. This greatly shaped, but also strengthened, his contributions to the Board of Longitude.

During his tenure, it increasingly collaborated with other influential organizations on a variety of pursuits, including the Admiralty and the Royal Navy, the Royal Observatory, the Royal Society, and corresponding foreign bodies. In addition to supporting Harrison's work, fostering the improvement of the lunar-distance method, and considering and encouraging other proposals, the Board expanded into the improvement of general navigation, science and technology.

For example, during and soon after Maskelyne's lifetime, it participated in the following: the development of more affordable marine chronometers than Harrison's 'one-offs'; the development of new and improved 'scientific' instruments; the trigonometric determination of longitude between different observatories and cities; organizing and staffing diverse voyages of discovery and 'science'; the search for a North-West Passage connecting the Atlantic and Pacific Oceans; outfitting the new Royal Observatory at the Cape of Good Hope and working with the Glass Committee, which searched for improved means of producing optical lenses for instruments such as telescopes. One can see in the surviving records how often the conduct of the earlier of these activities, and sometimes the Board's very involvement in them, was carried on the back of Maskelyne's existing interpersonal and institutional ties, and his activities at Greenwich.

Many of these activities were undertaken in partnership with the Royal Society, of which Maskelyne was an involved Fellow. He attended meetings when he could get away to do so, published more than fifty articles in its *Philosophical Transactions* and developed detailed instructions and increasingly standardized kits for the use of the observers on its expeditions. In 1775, the Fellows awarded him the prestigious Copley Medal, for his work on determining the Earth's density by using a plumb-line on Schiehallion mountain in Scotland. Bonds with such influential institutions, and with important individuals therein, also helped him to position the Board of Longitude as a prominent funding body and authority in navigation, science and technology – nationally and internationally (Fig. 4).

NATIONAL MARITIME MUSEUM, ZBA2362

Fig. 4: This silver medal was given to Maskelyne by the Institut National des Sciences et des Arts to mark his election as one of eight Foreign Associates in 1802. The medal, with a bust of Minerva, is by Rambert Dumarest and dated 1803.

This may not have been truly possible without Maskelyne's dedication and organization and, perhaps most importantly, without the wide-ranging networks of people and skills that he cultivated.

CULTIVATING 'SERVANTS' AND 'FRIENDS'

Maskelyne's working life was greatly defined by, and carried out through, his ever-expanding interpersonal networks. This was how most institutions and socio-economic relationships inherently operated in eighteenth- and early nineteenth-century Britain, so modern ideas of cronyism or nepotism are inapplicable. Trustworthiness and capability were often defined more by interpersonal connections and testimonials than they were by other forms of evidence including positions, publications, observations and inventions. This contributed to the frequency with which witnesses and collaborators became long-term associates of the Board of Longitude, with the same names reappearing over and over in the extant records. This could also blur or even erase the lines between employee and friend.

Literacy and style of communication of course also played some role in whether or not individuals were considered possible associates for institutions like the Board and the Royal Society. Artisanal, practical or intellectual knowledge, and skill alone did not always make a person suitable. For example, the prolific 'scientific' instrument seller, lecturer and publisher, Benjamin Martin (1704–82), tried for years to get himself elected to the Royal Society, with no success. In comparison, good interpersonal standing and esteem could help to outweigh a dearth of specific knowledge or expertise when it came to, for example, acting as a witness for longitude trials.

Maskelyne's interpersonal networks extended across Britain, much of Europe, and their current and former colonies. They encompassed every degree and combination of employee, colleague and friend – and in turn many of their associates. These networks reflected not only the astronomer's interests in mathematical and 'scientific' subjects but also his charitable contributions, social connections (including learned dining clubs) and family. Casual and long-term associates of all sorts were drawn from these overlapping socio-economic circles. They often styled themselves 'servants' or 'friends' of the astronomer, and sometimes received the same acknowledgment in return. The term 'friend' was somewhat more expansive during the eighteenth century than it is today, encompassing 'kinship ties, sentimental relationships, economic ties, occupational connections, intellectual and spiritual attachments, sociable networks, and political alliances'.[12]

A great part of Maskelyne's success in his different positions and pursuits lay in his apparent skill at establishing and maintaining such relationships, in person and through correspondence, and often accompanied by reciprocal kindnesses. We can trace many of these connections through surviving sources, including his own papers, now mainly held at the National Maritime Museum and in the archive of the Board of Longitude at Cambridge University Library. These are a rich trove of institutional and personal exchanges and announcements. The astronomer reused many pages as rough notepaper, so they are criss-crossed by later ideas, calculations and diagrams as well.

We can sample a cross-section of Maskelyne's intermingled personal and professional correspondence by looking at just one of the caches of these records.[13] This consists of forty-eight letters sent to him from 1765, the year he became Astronomer Royal, to 1809, two years before his death. They were sent from around the globe and touch upon: Maskelyne's work on astronomy both at home and for expeditions; the Board's publication of the annual *Nautical Almanac* and other texts; the lunar-distance and Jovian-moon methods of finding longitude at sea and on land; specific longitude projectors; and the use and complications of instruments such as the sextant.

Others are announcements seeking Maskelyne's presence at many of the institutions with which he was involved. For example, fellow Commissioner Lord Sandwich asked that he attend the election of a Professor of Chemistry at Cambridge in 1772, in the hope of swinging the vote towards a candidate from their *alma mater*, Trinity College. There are also a number of announcements of meetings of the Royal Society. Some letters describe the purchase of a pocket globe and of a telescope with a six-foot focal length from George and Peter Dollond, in 1807 and 1808. A small number address the workings of the Board of Longitude or the Royal Observatory. These include references to the Board's oversight committee reviewing accounts in 1798, an upcoming official visitation of the Royal Society to the Observatory in 1801, and samples of papers under consideration for use in publications in 1804.

Other letters in this volume mainly revolve around the astronomer's personal life and complement the personal manuscripts (including diaries and correspondence) which are now held at the NMM. In them we can see how interpersonal relationships and expressions of goodwill greased the cogs of Georgian society, business, and institutions. For example, in 1795 James Stuart Mackenzie, Lord Privy Seal of Scotland and a keen amateur astronomer, invited Maskelyne to dinner (Fig. 5).

He addressed the letter to 'My dear Tycho', in reference to the astronomer Tycho Brahe. He had originally sent the invitation to the Royal Society but it missed its recipient there.

The same year, Mr Rowed of the Globe Tavern in Fleet Street, London, wrote about an upcoming club meeting. In a

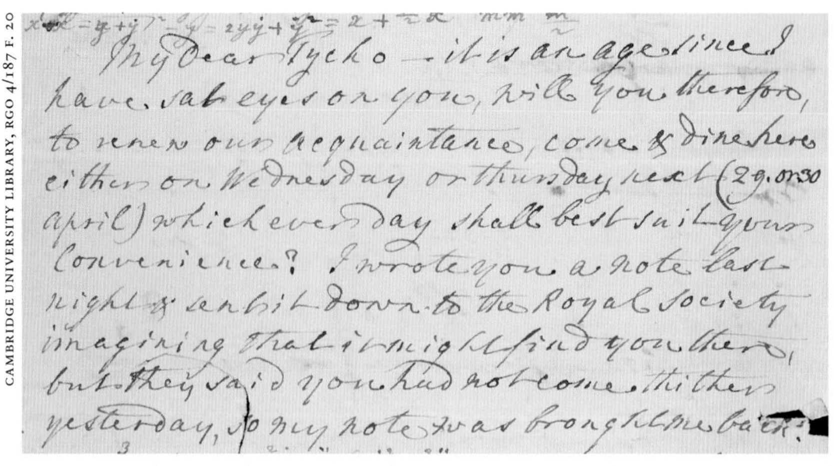
My Dear Tycho — it is an age since I
have sat eyes on you, will you therefore,
to renew our acquaintance, come & dine here
either on Wednesday or thursday next (29. or 30
april) whichever day shall best suit your
convenience? I wrote you a note last
night & sent it down to the Royal Society
imagining that it might find you there,
but they said you had not come thither
yesterday, so my note was brought me back.

Fig. 5: The salutation of a 1795 letter from James Stuart Mackenzie, a politician and amateur astronomer, calling Maskelyne 'My dear Tycho'.

letter of 1785, Charles Hutton, Professor of Mathematics at the Royal Military Academy, Woolwich, also mentioned a meeting of the club when he wrote about various subjects including *A Mathematical and Philosophical Dictionary*, upon which he was working. Hutton had been Foreign Secretary of the Royal Society until Joseph Banks organized his removal in 1783, which eventually led to his resignation and that of other mathematicians from the Society. In 1799, William Manwaring of Paddington, London, wrote about sending Maskelyne and his wife and daughter a present of pears and a pair of pigeons which he had bred.

Dr Layard, in an undated note, sought Maskelyne's vote for a candidate at the British Lying-in Hospital and thanked him for having obtained the proxy votes of his in-laws, the Ladies Clive (his sister, Margaret, having married Robert Clive). This was presumably Daniel Peter Layard, a successful man-midwife and physician. Other correspondents paid their compliments, some as friends seeking a meeting and some as strangers. As an example of the latter, teenager William Marsden (1754–1836) wrote from Fort Marlborough in Bencoolen in 1772, although the letter only reached Greenwich eight months later. Marsden apologized for approaching the astronomer without an introduction and then praised his *British Mariner's Guide* and method of finding longitude. He related his own observations made at sea with a sextant and concluded that, 'Any Service in my power on this Coast will be a pleasure to me to Perform you.' Marsden, then a junior clerk, advanced rapidly through the East India Company and later became a noted orientalist and numismatist – and a London friend of Maskelyne's from 1780 on.

Correspondents sometimes introduced others to the Astronomer Royal, as when, at the end of his letter of 1799, Thomas Wright wrote: 'I beg leave to introduce Mr. Webber to the honor of your correspondence; He is a person of Merit and of great application to the objects of his profession'. This was the American astronomer Samuel Webber, who had recently assisted Wright, the Surveyor-General of the colony of St John's Island, in taking accurate sightings to establish the positions of the various rivers then claimed to be the border between New Brunswick and the District of Maine in Massachusetts.

The correspondents represented in this specific volume are mainly British, including those posted or travelling abroad, but include a small number of contacts from foreign nations. (A large and diverse host of foreign correspondents can be found within other collections of Maskelyne's letters.) The Britons here mainly included astronomers and computers, and members of the government and naval and military establishments – including the soon-to-be-deposed Governor William Bligh of New South Wales, earlier of the *Bounty*. Most of these correspondents were male, with Mary Edwards of Ludlow being a notable exception, as covered in Chapter 4 of this book. In 1785, a year after being widowed, Edwards wrote to direct Maskelyne to settle her late husband's accounts and to thank him for his assistance.

On the foreign side, two officers of the War Ministry in Madrid, in Napoleonic Spain, conveyed observations of a solar eclipse made by Ali-Beik Abd-Allah at Tangier in 1803. Three years later, Julian Canelas wrote from an observatory on the Isla de León (between the peninsular city of Cádiz and the Spanish mainland) to ask for observations of a recent solar eclipse and of the occultations of Antares to compare with his own. Gian Giuseppe Barzellini sent a paper from the Austrian Habsburg city of Gorizia about his calculations for a solar clock in 1774 (Fig. 6).

Barzellini, whose papers and books are now held at the theological seminary in Gorizia, was the first Director of its insurance bank, worked with the local agrarian reform society, and conducted a land survey of the province. He said in the letter that he taught himself mathematics, astronomy and instrument design

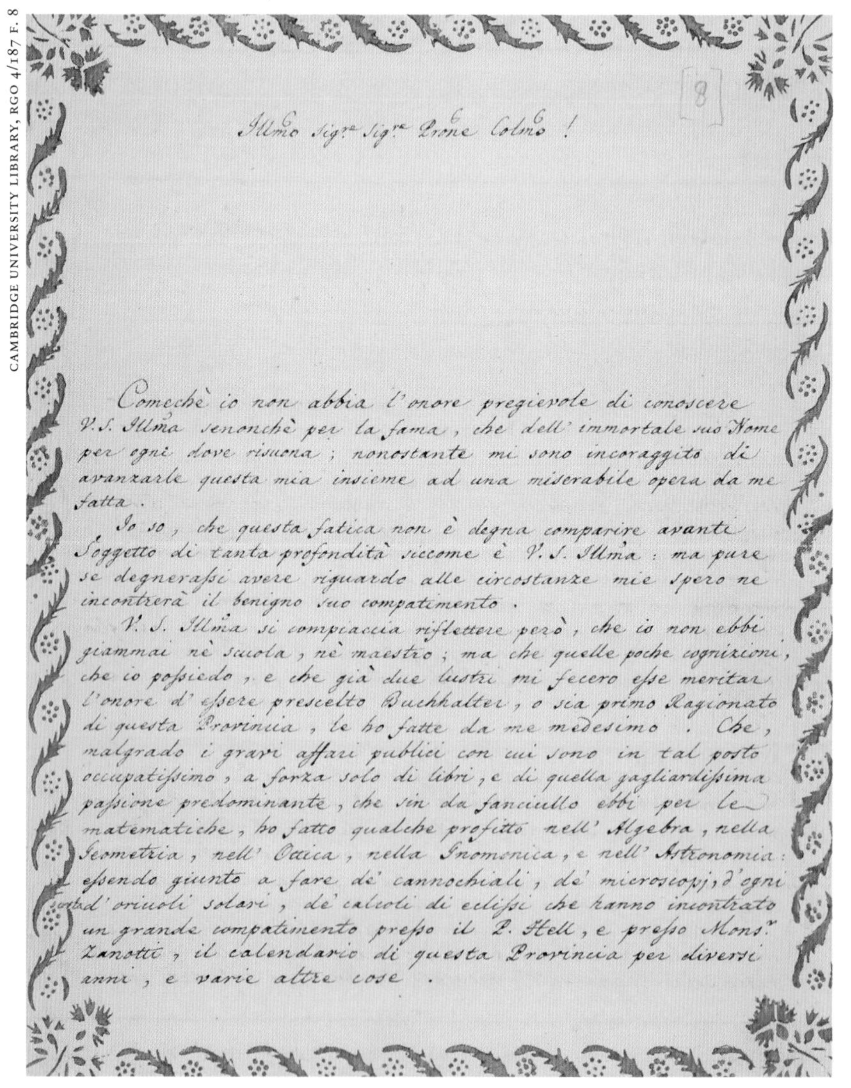

[8]

Illmo Sigr. Sigr. Prone Colmo !

Comechè io non abbia l'onore pregievole di conoscere V. S. Illma senonchè per la fama, che dell' immortale suo Nome per ogni dove risuona; nonostante mi sono incoraggito di avanzarle questa mia insieme ad una miserabile opera da me fatta.

Io so, che questa fatica non è degna comparire avanti Soggetto di tanta profondità siccome è V. S. Illma: ma pure se degnerassi avere riguardo alle circostanze mie spero ne incontrerà il benigno suo compatimento.

V. S. Illma si compiaccia riflettere però, che io non ebbi giammai nè scuola, nè maestro; ma che quelle poche cognizioni, che io possiedo, e che già due lustri mi fecero esse meritar l'onore d' essere prescelto Buchhalter, o sia primo Ragionato di questa Provincia, le ho fatte da me medesimo. Che, malgrado i gravi affari publici con cui sono in tal posto occupatissimo, a forza solo di libri, e di quella gagliardissima passione predominante, che sin da fanciullo ebbi per le matematiche, ho fatto qualche profitto nell' Algebra, nella Geometria, nell' Ottica, nella Gnomonica, e nell' Astronomia: essendo giunto a fare de' cannochiali, de' microscopj, d'ogni sorta d'oriuoli solari, de' calcoli di eclissi che hanno incontrato un grande compatimento presso il P. Hell, e presso Monsr. Zanotti, il calendario di questa Provincia per diversi anni, e varie altre cose.

Fig. 6: A letter from Gian Giuseppe Barzellini to Maskelyne, sent in 1774, containing calculations relating to a solar clock.

by studying publications. Four years after this, he constructed a meridian line on the exterior wall of the cathedral. It was not unusual for such partially or wholly self-taught enthusiasts to contact Maskelyne for his opinion or to try to contribute to larger astronomical efforts.

Of perhaps the greatest interest to *aficionados* of longitude, and of the story of the clockmaker John Harrison, is an undated draft letter of 1773 in this volume. Maskelyne sent it to Lord Sandwich, who was First Lord of the Admiralty and thus the leading Commissioner. In it, the Astronomer Royal first responds to Harrison's applications to the Board of Longitude and to Parliament to receive a second £10,000 in reward money for his marine timekeepers, despite refusing to produce two more watches and otherwise submit to the trials required by the most recent legislation. The astronomer then asks leave to provide his correspondent with an abstract and explanation of the Act of 1714, which he had at first intended

> as an introduction to an answer to Mr. Harrison's scandalous [originally drafted as 'abusive'] pamphlet on my trial of his watch at the Royal Observatory; tho I afterwards dropt the design of publishing the same thinking such abuse thrown out without probability or proof required no refutation.[14]

As we have seen elsewhere in this book, Maskelyne never responded in published form to the slights from John and his son William, put forth in the 1765 pamphlet *A Narrative of the Proceedings relative to the Discovery of the Longitude at Sea* and in newspapers and journals – although anonymous supporters on either his or the Board's side did rally in the media.

Of course, despite his apparent skill at building and maintaining interpersonal relationships, Maskelyne could not entirely avoid interpersonal conflict – most famously with the Harrisons. However, as has already been noted, the same tensions are likely to have arisen had any Astronomer Royal been pursuing a 'competing' longitude method, and especially before the time of his appointment. The situation was aggravated by the institutionalization (or bureaucratization) of the Board of Longitude to which Maskelyne contributed, and which had somewhat ironically been initiated by Harrison's early successes. These developments slowed the operations of the Board, which had previously granted repeated financial and legislative encouragement to the clockmaker with apparent ease. They also contributed to greater thought being put into the definition of the Board's purpose and of the terms for its rewards. Beyond the Harrisons, Maskelyne also fell out with Joseph Banks at least twice (as is discussed in Chapter 7), although they worked and sometimes socialized together in different capacities for about forty years. However, the surviving evidence suggests that Maskelyne successfully conducted most of his myriad relationships with everything from polite professionalism to close and collaborative friendship.

THE LOYALTIES OF A LIFETIME

Many luminaries across Europe praised Maskelyne warmly after meeting or corresponding with him. Delambre recalled that, 'Every astronomer, every learned man, found in him a brother'.[15] Joseph Bernard, Marquis de Chabert – who sat on the Bureau des longitudes – praised his tact and kindness in dealing with French refugees to Britain. In 1770, in his three-volume *Londres*, Pierre-Jean Grosley wrote that Maskelyne was a man 'in whose company I found a politeness and indulgence towards passing visitors that scholars of this rank do not always have'.[16] The pioneering female astronomer, Caroline Herschel, commemorated her warm friendship and collaboration with Maskelyne in the late eighteenth and early nineteenth centuries. The Astronomer Royal appreciated Caroline's assistance and accomplishments, repeatedly invited her to Greenwich, saw that her star catalogue and corrections to Flamsteed's observations were published by the Royal Society, and gave her a pair of binoculars and a night-glass.

Many of Maskelyne's social and working relationships extended across decades, and this encompassed the lesser-known participants in his personal and institutional activities as well as the Lalandes, Bankses and Herschels. For example, he corresponded with the Scottish astronomer, Andrew Mackay, from at least the late 1780s to the early 1800s, as we can see from surviving correspondence at Cambridge and Greenwich. Mackay was then the unpaid head of the observatory at Marischal College in Aberdeen and was partially or wholly self-taught. The association between the two astronomers began when Mackay asked Maskelyne in the spring of 1787 to lay his 'new method of finding the Longitude and Latitude of a Ship at Sea' before the Board of Longitude, which was done on 8 December. Maskelyne was so impressed by the Scotsman's treatise, position in Aberdeen, and apparent skills, that he suggested bringing him to Greenwich as an assistant. The treatise in question was later published by subscription as *The Theory and Practice of finding the Longitude at Sea or on Land*, and various editions were dedicated to the Astronomer Royal, although the Board declined to give its author an award.

Maskelyne, normally so adept at navigating the steps of Georgian social interaction, appears to have made a slight *faux pas* here by not having first considered Mackay's connections higher up the social and collegiate ladder. He wrote in an ensuing letter that he would not want to steal the Scotsman away from his patrons at Marischal College and especially from Professor Patrick Copland (an associate from the Schiehallion days) without their approval. The two men continued to discuss the details of the assistantship, but it did not come to pass. During the ensuing years, Maskelyne and Mackay appear to have struck up one of the former's common mentoring friendships. The two men exchanged greetings and astronomical and mathematical information, with the Astronomer Royal sometimes sending books to Mackay. On 2 August 1790, the elder responded positively to the Scotsman, having attempted to see him at Greenwich while he was visiting England.

Later in 1802, Maskelyne tried to make his young friend the replacement astronomer on the Matthew Flinders voyage to Australia, despite confusion over the pay being offered by the

Board of Longitude and stalling on the part of the supposed supplicant, who mainly hoped to win a professorship at King's College in Aberdeen. The surviving letters show that, at this time, Maskelyne was also keen on helping Mackay to become a Fellow of the Royal Society in London, provided he wanted to go to the significant expense involved. He offered to try to secure the support of the President, Joseph Banks, after which he said Mackay's election would be practically assured. The two astronomers discussed possible co-signers for the election certificate, with the Astronomer Royal expressing certainty of the Spanish-born Joseph de Mendoza y Rios but alerting Mackay that the other individual he suggested had not attended meetings since the blow-up between Banks and the mathematical Fellows in the early 1780s. He instead proposed the Scotsman's friend, Professor Abraham Robertson of Oxford, or others including William Herschel and Alexander Aubert.

Also, later in 1802, Maskelyne regretfully informed Mackay that he had chosen someone else to join the Flinders expedition, as it could not be put off any longer. He added that the Scotsman would not have been able to go anyway if his litigation, begun against King's College after having been passed over for the professorship, were successful. However, he assured his young friend that he would still support his election to the Royal Society, and even that he would try to get his longitude book 'advertised on the sheets of the nautical almanac'. He went on to encourage Mackay's interest in naval architecture and asked to be put down as a subscriber to his new treatises on astronomy and navigation. The Scotsman moved to London in 1804 and, in addition to teaching there, was an examiner in mathematics for Trinity House, Christ's Hospital and the East India Company. In 1805, Maskelyne was still discussing getting him elected to the Royal Society, although this never happened, and communicating favourably about the third edition of his longitude book.

Nevil Maskelyne continued a multitude of correspondences like this, so central to the operation of his career as Commissioner and as an astronomer and natural philosopher, until his death in 1811 at the age of seventy-nine. It was fitting that one of the last people with whom he communicated was Henry Andrews, a computer for the *Nautical Almanac*, introduced in Chapter 4. Letters now at Cambridge reveal a cordial working relationship between the two men, with occasional evidence of private as well as professional ties.[17] In December 1788, Maskelyne reported that he and his wife were soon to interview a prospective nursery maid for their three-year-old daughter, Margaret, so that Mrs Andrews should no longer trouble herself to try to find one. The astronomer related in the next letter that they had taken up a suggested female servant, but that she had behaved badly and left, and later tried to impose upon Mrs Maskelyne for references.

In 1810, the astronomer sent Andrews a twelve-inch celestial globe with a quadrant of altitude, and an accompanying book. This was to help the computer to work with lunar eclipses by marking the Moon's location on the globe with red ink, which could then be mostly rubbed off with a cloth. Maskelyne gave the same pricey gift to three other computers as well, including the similarly long-serving Mary Edwards. The Astronomer Royal gradually came to consult with Andrews more openly on astronomical matters, such as on the predicted appearance of a comet in 1790. He also educated the computer about new developments in their field, including an innovation made by Pierre-Simon, Marquis de Laplace, in 1798. In 1793, when Andrews was to make one of his rare trips to Greenwich, Maskelyne hoped to dine with him and promised to try to find a position for his son. Later that year, he warned him that he might not have more *Nautical Almanac* work for five years, as his ongoing work and recent publications by Lalande needed to be incorporated first. However, he and the Board did their best to make work for the computers, who expressed fears about their livelihoods in the interval, and set them to redoing earlier almanacs in the light of Lalande's and Charles Mason's new tables.

Maskelyne continued to work with Andrews until the end of

his life. Touchingly, it was his by-then adult daughter, Margaret, who wrote on 7 February 1811 to notify the long-time computer, at her father's behest, that he was very ill and that work on the *Nautical Almanac* would have to stand still until he was better. Two days later, the astronomer passed away at the Royal Observatory which he had served for forty-six years. The public consensus was, as *The Gentleman's Magazine* said, that 'not only his own, but every other country bears witness, that few have fulfilled those duties with so much ability, none perhaps with so much utility to the publick'.[18]

Andrews continued working as a computer for the next Astronomer Royal, John Pond, until his own death four years later. Pond was a highly regarded practical astronomer who instituted observational and technological innovations at Greenwich. However, he let the administrative side of the position slip, including the production of the annual *Nautical Almanac* and the issuing of marine chronometers to the Royal Navy. He also did not establish networks of communication quite as ambitious or as wide-ranging as those of his predecessor. It was not primarily through his astronomical skill that Nevil Maskelyne had forged a new future for the Board of Longitude and for maritime and scientific endeavour in Britain: he mainly did so through the cultivation of interpersonal and institutional ties that spanned the globe and through the application of an admirable amount of vision, dedication and professionalism to his responsibilities.

FURTHER READING

In addition to the sources already cited by my fellow authors and the digitized Board of Longitude archives (http://cudl.lib.cam.ac.uk/collections/longitude), there are many publications related to the broader Georgian dynamics which I discuss.

For example, much has been written (and debated) about the early modern concept of 'politeness' since at least the late 1980s. The term represents a somewhat ambiguous set of hallmarks for the well-bred gentleman or lady, including proper and tasteful behaviour, interpersonal interactions, material display, and learning. This sometimes-vague framework was important for the practice and success of the personal interactions and relationships which dominated the activities and resources of Maskelyne and of the Board of Longitude at large. The sources available on the subject include: Paul Langford, 'The Uses of Eighteenth-Century Politeness', *Transactions of the Royal Historical Society*, 12 (2002), pp. 311–31; Lawrence E. Klein, 'Politeness and the Interpretation of the British Eighteenth Century', *The Historical Journal*, 45 (2002), pp. 869–98; Helen Berry, 'Polite consumption: shopping in eighteenth-century England', *Transactions of the Royal Historical Society*, 12 (2002), pp. 375–94; and Woodruff D. Smith, *Consumption and the Making of Respectability, 1600–1800* (Routledge: London, 2002).

There are many other publications about the nature and dynamics of interpersonal relationships during the early modern period as well – either broadly focused, or in the form of important case studies like that of the Verney family. Examples include: Helen Berry and Elizabeth Foyster (eds), *The Family in Early Modern England* (Cambridge University Press: Cambridge, 2011); Naomi Tadmor, *Family and Friends in Eighteenth-Century England: Household, Kinship, and Patronage* (Cambridge University Press: Cambridge, 2001); and Susan Whyman, *Sociability and Power in Late-Stuart England: The Cultural Worlds of the Verneys 1660–1720* (Oxford University Press: Oxford, 1999).

Other sources help to illuminate the sorts of Georgian learned and social organizations to which Maskelyne and many of his associates belonged. The activities and resources of institutions such as the Board of Longitude were often shaped by or even carried out in such spheres as well as in official meetings, given their overlapping memberships and the highly interpersonal conduct of early modern institutions. Relevant examples include: Trevor H. Levere and Gerard L. E. Turner, *Discussing Chemistry and Steam: The Minutes of a Coffee House Philosophical Society 1780–1787* (Oxford University Press: Oxford, 2002); Jenny Uglow, *The Lunar Men: The Friends who Made the Future*

(Faber: London, 2002); Richard Sorrenson, 'Towards a History of the Royal Society in the Eighteenth Century', *Notes and Records of the Royal Society,* 50 (1996) pp. 29–46; Michael Hunter, *The Royal Society and its Fellows, 1660–1700: The morphology of an early scientific institution*, 2nd edn (British Society for the History of Science: Chalfont St Giles, 1994); and many of the chapters in Larry Stewart, *The Rise of Public Science: Rhetoric, Technology, and Natural Philosophy in Newtonian Britain, 1660–1750* (Cambridge University Press: Cambridge, 1992).

Finally, there are also resources relating to the institutions, including the Admiralty and Royal Navy, that may have influenced and encouraged the increasing institutionalization and bureaucratization of the Board of Longitude, which began in the 1760s, such as: Roger Morriss, *The Foundations of British Maritime Ascendancy: Resources, Logistics and the State, 1755–1815* (Cambridge University Press: Cambridge, 2011); Clive Wilkinson, *The British Navy and the State in the Eighteenth Century* (Boydell Press: Rochester, NY, 2004); and Gloria Clifton, 'The London Mathematical Instrument Makers and the British Navy, 1700–1850', in Pieter van der Merwe (ed.), *Science and the French and British Navies, 1700–1850* (National Maritime Museum: London, 2003), pp. 24–33.

CASE STUDY F

THE ROYAL SOCIETY AND GEORGIAN SCIENCE

As Astronomer Royal, Maskelyne was inevitably a key figure within the Royal Society and, more broadly, a representative of scientific knowledge and interests. As several of the chapters in this book have revealed, he was considerably more active in this respect than his immediate predecessors and more akin to the first two holders of the post, John Flamsteed and Edmond Halley. Like Maskelyne, both these men conducted wide national and international correspondence, acted as advisers in all sorts of situations when required and were active in the Royal Society – at least, in Flamsteed's case, until his rupture with Isaac Newton. By Maskelyne's time, after a period of financial difficulty, the Society was firmly established, and although it still had very much of the character of a gentleman's club, it was increasingly accepted as a place of independent authority and an advisory body to the government.

This was, in part, present from the Society's foundation. Inspired by the inductive empiricism of Francis Bacon (1561–1626), it aimed at an alliance between philosophers and politicians that might put the 'new philosophy' to use and improve mankind's general lot. The early Society was full of politicians, courtiers and diplomats but, as yet, the government lacked a sufficient system of administration and civil service to make use of the expertise potentially available. The need for a bureaucracy to do this was underlined by the fact that, where the relationship between the State and science did blossom, it was through the large military departments, particularly the Admiralty.[1] Figures like Jonas Moore, Master-General of the Ordnance and a crucial figure in the foundation of the Royal Observatory, and Samuel Pepys, Secretary of the Admiralty, were early examples of the links that also led to the Longitude

Act of 1714, which ultimately united politicians, officials from the Admiralty and men of science within a standing body.

Changes in government structure, and the increase of spending associated with an enlarged civil service, came in the reign of George III (1738–1820), Maskelyne's near contemporary. They were the inevitable concomitant of rising military, trade and imperial competition with European powers, especially France, which had a much more centralized government that was already better able to benefit from the work of its savants. It was on the rising tide of this competition that Maskelyne was able to launch himself into significance in the scientific world by pressing for Britain to undertake its own expeditions to view the transit of Venus. His additional observing plans for his voyage to St Helena were also designed to show that the British could compete in such ambitious overseas projects with Continental astronomers and geodesists. Where British science increasingly had a lead was in being able to benefit from its internationally recognized expertise and skill in the manufacture of scientific instruments, facilitated by the open systems of commerce and scientific discussion in London. These elements were particularly significant for the development of research into longitude schemes.

Britain's rise to maritime dominance, in trade and military spheres, was of huge importance to the development of British science. It allowed access to new and distant places for the collection of artefacts, specimens and physical observations. It was Joseph Banks, the wealthy and privately funded botanist on Cook's first voyage, who was able to take greatest advantage of the new order. He had decided to devote himself to 'the Scientific Service of the Public' and took it on himself to show government where science could serve their needs, and to highlight how its cultivation had clear utilitarian ends.[2] During Maskelyne's term as Astronomer Royal, from 1765 on, much of this endeavour centred around Banks' involvement with voyages of exploration, collecting samples and the foundation of Kew Gardens, all aimed at gaining botanical and other scientific knowledge that might help exploit new natural resources and support the growth of Britain's trade and empire. His influence also later reached into a host of new organizations, including the Linnean Society, Horticultural Society, Royal Institution, British Museum, Society of Dilettanti and Board of Agriculture.

Despite its focus on utility and science's potential to support the State, the Royal Society distinguished itself from some of these other bodies and those, like the Society of Arts, that offered financial incentives for specified improvements or new ideas. It embodied an ethos of disinterestedness which meant that those with careers in fields like mathematical teaching or instrument manufacture had to present their work as within the framework of natural philosophy, conducted for the general public good rather than individual gain. This is one of the elements behind the clash that took place between the members of the Board of Longitude and John Harrison, the former adhering to an ideal of scientific work that involved open communication, and the latter wanting to protect his commercial advantage. Likewise, engineers and industrialists who wished to be Fellows of the Royal Society needed to find ways to repackage their professional and practical work as generalized investigations into the workings of nature.

In the last quarter of the eighteenth century, the Society and, above all, Banks as its President, were to be engaged in an increasing number of projects that involved government expenditure. These included a wide range of special committees of enquiry on topics like provision of hemp for the Navy, the safety of gas lighting, the properties of particular substances or the best materials for specific uses. Most ambitious, after the transit expeditions, was the geodetic survey of England – the Ordnance Survey – supported with a £3,000 royal grant, which included the attempt to establish the difference in longitude between the Greenwich and Paris observatories and to link the English with the French survey. Here, as with Banks's close involvement in instructing and selecting the non-astronomical scientific personnel for voyages of exploration, there was always potential cross-over or rivalry in the interests and expertise of the Royal Society, its President, the Astronomer Royal and the Board of Longitude.

Within the Royal Society, at least, Banks was able to gain a firm grip on power after being elected to the presidency in 1778. This was challenged in the 1780s in what became known as the Society's 'Dissensions'. As so often, these disputes had several characteristics and represented a number of different views. In one aspect they were the final consolidation of Banks's authority, precipitated as they were in part by the removal as Secretary of Paul Maty, a representative of the group that had previously been dominant in the Society, known as the 'Hardwicke circle'.[3] Opposition to this move related for some to older alliances, while for others it was an objection to the autocratic nature of what has become known as the 'Banksian regime', which stated that only the President and Council of the Society, not its members, could select officers and, effectively, other council members. In some aspects they represented divisions within the Society's membership relating to class, or to views about what kind of individual should represent science.

The 1784 dissensions have also been dubbed the 'Mathematicians' Mutiny'. As well as Maty, the mathematician Charles Hutton had been forced to resign as the Society's Foreign Secretary. This led to a threat by Hutton's supporters, including the mathematicians George Atwood, Francis Maseres, Samuel Hornsby, John Landen and Maskelyne, to secede. They saw themselves as standing for 'professional skill', against the gentlemanly dilettantism of some of Banks's circle, and for 'that accuracy of science which arises from having been employed only on one subject'.[4] Their secession, it was claimed, would leave Banks with his 'train of feeble *amateurs*'.[5] These amateurs, however, suspected a 'levelling spirit' from some of the mutineers and maintained that science and the Royal Society could best be guided by the landed classes, whose decisions would be in the interest of the nation in which they had a stake, and by men whose learning and experience afforded them a broad and disinterested perspective, unavailable to those trained only in one discipline to which they might also owe their living.[6]

While Hutton was evidently tiresome and dispensable to Banks, he and Maskelyne were personally close. Maskelyne had spoken up for Hutton at the Council meeting that charged him with neglect of his position, asking that he should at least be able to answer the charges before any decision. When later speaking in defence of the 'remarkably diligent' Hutton, Maskelyne apologized 'for taking up the time of the Society' and for 'being unaccustomed to speak in public'. He was sorry, too, 'that the necessary defence of a worthy member might throw blame in another quarter' but stated that 'a regard to justice impelled him to it'. He undoubtedly felt that he must support his friend, a man whom he called:

> one of the most learned and deserving of their members, who has communicated many valuable papers to them and is likely to communicate many more; who is indefatigable in business, has spent his whole time in the promotion of science; and is one of the greatest mathematicians in Europe.[7]

The details of these events and their outcome are the subject of the following chapter. As it shows, however, it was also necessary for Maskelyne to find ways to continue working with Banks.

In 1811, it was the Board of Visitors to the Royal Observatory that recommended Maskelyne's successor. This was made up of the President, Vice-President and members of the Council of the Royal Society, and so John Pond was really Banks' appointment. As a man of very different character and abilities from Maskelyne, his accession represented an even greater concentration of power for the Banksian regime. This had by then come to dominate the Society and had extended its influence into the scientific interests of the Admiralty, including the Royal Observatory and the Board of Longitude. The result, in addition to Pond's appointment, was the Observatory's subordination to the Admiralty from 1816 and changes to the Board of Longitude, particularly after it was reconstituted in 1818. These changes led to dilution of the influence of the mathematical and astronomical professors who had been *ex officio* Commissioners of Longitude since 1714. Now these troublesome mathematicians were balanced by three Fellows of the Royal Society, selected by its President and Council, and three salaried Resident Commissioners, selected by the Board as a whole. It was not long, however, before astronomers founded their own society, which, if not a secession from the Royal Society, was to form a significant locus for criticism of the Banksian regime and its legacy. It was also, ultimately, a significant locus of expertise equal to the Society in matters such as the publication of the *Nautical Almanac* and oversight of the Royal Observatory.

Beyond the power struggles of boards and societies, science was gaining ever-wider interest and recognition. Books and lectures on natural philosophy, accounts of voyages of exploration, astronomical diagrams and instruments, and chemical, magnetic and electrical demonstrations were becoming ever-more popular with increasingly broad audiences. There was real public interest in the return of Halley's Comet in 1759, as confirming the power of Newton's theory of gravitation. The transits of Venus also sparked broad interest (prompting George III to build his own observatory at Kew), as did the discovery of a new planet (Uranus) by William Herschel in 1781. The prestige of scientific eminence, both for the nation as it competed with France and other European powers, and for leaders of the scientific world like Banks and Maskelyne, was increasingly significant. Nevertheless, the age of the scientist – of regular career paths, formal education and training, and of large-scale government investment – was still a very long way off: so, for the time being, personal ties, patronage, correspondence, opportunism and luck remained essential to those intent on making a living from science.

7

FRIEND AND FOE: THE TEMPESTUOUS RELATIONSHIP BETWEEN NEVIL MASKELYNE AND JOSEPH BANKS

Caitlin Homes

'After some early disagreements, [Joseph Banks] became the close friend of the Astronomer Royal, Nevil Maskelyne.'[1]

THE INTRODUCTIONS

In the latter part of the eighteenth century, a talented, charismatic and relatively young man took up the chair of the President of the Royal Society. At the age of thirty-five, Joseph Banks (1743–1820) was to become the longest-serving President in its 350-year history, holding the prestigious title for just over forty years, from November 1778 until his death. Before doing so, however, he had to survive a short-lived but determined attack from a small group of Fellows sometimes described as 'the Mathematical Faction'. This conflict, which became known as the 'Dissensions', took place between November 1783 and May 1784, five years into Banks's presidency.

When the key issues of the dispute were finally decided, it was clear that Banks was victorious and his authority more entrenched than ever. Most of the handful of men who had taken a public stand against him effectively disappeared from the Royal Society, more or less quietly. The most noteworthy Fellow remaining was Nevil Maskelyne who, by virtue of his office as Astronomer Royal, could resign neither from his post nor the Royal Society without completely sacrificing his scientific career. Thus, across three decades, the two men were forced by their respective roles to rebuild a close working relationship, particularly within the Board of Longitude, as well as maintain some level of social accord within the Society. The general gloss on their story is often expressed by historians in a similar fashion to Richard Holmes's description, quoted above, but the tangled fluctuations of friendship and conflict between them were more complex and interesting.

Maskelyne and Banks were by no means natural enemies, although there were some marked differences between them. Banks was eleven years younger, the eldest son of a wealthy land-owning family. He attended Eton and Oxford University, and was able to make up for Oxford's lack of provision of lectures in botany, his great passion, by paying for a tutor from Cambridge to come across to run courses. On reaching his majority he left Oxford without a degree, as such a qualification was unnecessary to a gentleman with his resources and social status.

In contrast, Maskelyne had been orphaned at the age of fourteen, and as a third son he received little in the way of inheritance. After his education at Westminster School he attended Cambridge University as a sizar, undertaking menial tasks in return for reduced tuition fees. Maskelyne took the Mathematical Tripos with some distinction in 1754 and was then ordained, one of the requirements for being elected a Fellow at Trinity College, which provided him with a small salary. He worked as a curate between 1755 and 1763, occasionally officiating at weddings, and even after being appointed Astronomer Royal he held stipends from several parishes during his lifetime, which were not insignificant contributions to his income. He also took the degree of Doctor of Divinity in 1777, although there is little written evidence to indicate the extent of his personal faith. Had he not become Astronomer Royal, he would most likely have had a career in the Church and would have had to maintain his scientific interests as a keen hobby.

Banks, on the other hand, shared neither Maskelyne's interest in religion nor his need for a salary: he was, if anything, dismissive of the Church and clergy, while his independent wealth enabled him to make his botanical enthusiasm and his role as President of the Royal Society the centre of his life, without recourse to anybody else. However, at the time of their first meeting, probably in 1767 or 1768, these two men had a major life experience in common that was shared by few others; that of the sea-faring adventurer in pursuit of scientific knowledge. The first of Maskelyne's voyages was undertaken at the age of twenty-eight while Banks was still at Eton. When he was appointed Astronomer Royal, four years later, he exchanged his sea-faring days for a prestigious lifelong career that enabled him to pursue his scientific interests full-time, and also gained the authority to advise on the appointment of other astronomers to accompany the many voyages of discovery that came to characterize the period.

At about this time Joseph Banks had moved to London and was working alongside naturalist Daniel Solander in the British Museum. He was just twenty-three when he embarked on a self-funded, independently directed course of field work in Labrador and Newfoundland, in 1766. Over a period of nine months he collected plant and animal specimens, observed geological strata and gathered information about the native 'Indians'. He was made a Fellow of the Royal Society in his absence and first attended meetings upon his return in February 1767, where he most probably met Maskelyne during the following months. They already shared an important common acquaintance: Israel Lyons, Banks's botany tutor in 1764, was one of the first computers engaged to work on the *Nautical Almanac* by Maskelyne in 1765.[2]

A year later their paths crossed more significantly in preparation for perhaps the greatest event in Banks's life – accompanying (then Lieutenant) James Cook on his first voyage to the South Pacific to observe the transit of Venus in 1769. As we have already seen in Case study E, Maskelyne, as the Astronomer Royal, was the major scientific adviser to this and the four other British expeditions, sent off to observe the transit at home and abroad. Banks saw the remarkable opportunity before him and negotiated persuasively with the Royal Society and the Admiralty, convincing them to allow him enough space on Cook's *Endeavour* to take on board his team of eight men and equipment, making the study of natural history one of the goals of the voyage and ultimately one of its greatest triumphs. One of the reasons for Banks's successful negotiations was his ability to finance this extra part of the venture himself, making no demands on either the Admiralty (which did contribute his party's general victualling) or the funds of the Royal Society as provided by King George III.

One of the earliest pieces of direct evidence of Maskelyne and Banks meeting is found on 5 May 1768, the day Cook was introduced to the Royal Society Council and accepted their offer to be an observer of the transit in Tahiti. Their names are recorded as attending the Society's dining club at the Mitre tavern, Maskelyne as a member and Banks as a guest.[3] Banks was yet to be admitted as a member of this exclusive gathering, which met for dinner prior to meetings of the Society, and which was known officially as the Club of the Royal Philosophers and, more informally, as 'The Mitre Club'. Banks attended the dining club three times in February that year, shortly after the announcement of the *Endeavour* voyage, and no doubt used these opportunities to discuss the possibility of including a party of natural historians. He was given a great honour by the Club, being made a member in his absence in 1770, and upon the *Endeavour*'s return in July 1771 dined with it on the 25th of that month, no doubt to the great entertainment of the fifteen other Fellows present, Maskelyne included.

In these early days, Maskelyne, Astronomer Royal and eleven years senior to Banks, would have had considerable status in their relationship. Indeed, Banks refers to him in his account of St Helena (in his *Endeavour* journal), acknowledging that he was walking where Maskelyne had been before him (Fig. 1):

> Spent this day in Botanizing on the Ridge where the Cabbage trees grow, visiting Cuckolds point and Dianas peak, the Highest land

in the Island as settled by the Observations of Mr Maskelyne, who was sent out to this Island by the Royal Society for the Purpose of Observing the transit of Venus in the Year [1761].[4]

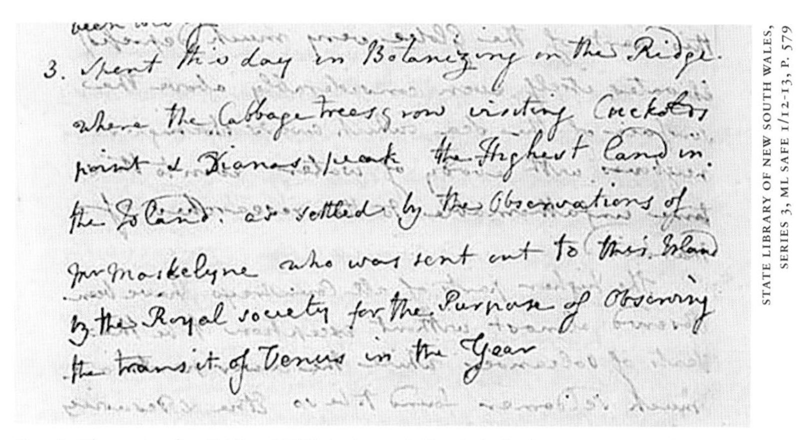

3. Spent this day in Botanizing on the Ridge
where the Cabbage trees grow visiting Cuckolds
point & Dianas peak the Highest Land in
the Island: as settled by the Observations of
Mr Maskelyne who was sent out to this Island
by the Royal society for the Purpose of Observing
the transit of Venus in the Year

Fig. 1: The entry for 3 May 1771 in Joseph Banks's *Endeavour* journal, written while at the island of St Helena.

STATE LIBRARY OF NEW SOUTH WALES, SERIES 3, ML SAFE 1/12-13, P. 579

Even Banks's close identification with Cook's successful and famous voyage (see, for example, the artefacts and Maori cloak in Banks's portrait, Fig. 2) would have reflected back on Maskelyne as someone of authority in planning the expedition, and was not something that caused any threat to his position.

Rather, as will be seen from the letter quoted below, he seems to have delighted in the younger man's adventures and scientific enterprise, even though Banks's interests were in a vastly different field from his own, and he was keen to promote Banks's involvement with the Royal Society as much as he could.

Following the *Endeavour* voyage, Banks made a final journey of scientific exploration to Iceland with a team of sixteen. This was a last-minute compromise after he had proved too demanding in his requirements to join Cook's second voyage to the Pacific and abandoned the attempt, walking off after the Navy ordered the adaptations required to accommodate his party removed from the *Resolution* for reasons of safety. Upon his return to London, Banks's involvement with the Royal Society increased significantly: he was appointed to the Council in November 1773,

Fig. 2: Portrait of Joseph Banks, by Benjamin West, painted after his return from the circumnavigation of the *Endeavour* in 1771. He is surrounded by items collected during the voyage.

became one of the official Visitors to the Royal Observatory, and was on the 'Committee of Attraction' that requested Maskelyne to undertake one further field trip himself in 1774. This, as we have previously seen, was Maskelyne's four-month foray to Scotland to undertake gravitational experiments on Schiehallion, for which the King gave him leave from the Royal Observatory.

In the surviving correspondence between the two men, the earliest letter is also the warmest. Written in October 1775 from Maskelyne to Banks, it opens with thanks for the latter's information on the rocks that Maskelyne had brought back from Scotland: 'I am very happy in finding your opinion so clearly in favor of Schehalliens being a Virgin Venus never submitted to the embraces of Vulcan, the God of Fire: the more estimable, & more attractive on that account!' It continues in a chatty, friendly tone and finishes with a confident flourish: 'I shall do myself the pleasure to call on you the first morning in my power, & shall beg a sight of your Chameleon' (Fig. 3).[5]

In both of these sentences are glimpses of who these men are and how they relate to each other: Nevil Maskelyne, the current year's winner of the Copley medal for 'weighing the world' at Schiehallion, and Joseph Banks, the famous collector of new and extraordinary flora and fauna, are sharing a mildly risqué joke! One is happily expecting a warm welcome and intriguing conversation with the other at the soonest opportunity. Banks's hospitality to gentlemen of science throughout his life is well known and, in the early years of their acquaintance, Maskelyne felt confident to accept and enjoy it.

THE 'REBELLIOUS' DINING CLUB

This apparent camaraderie between Maskelyne and Banks was not to last; or, at least, there is no other surviving letter between them that evokes such warmth and affection. A series of events over the next ten years, culminating in the Dissensions, created a barrier between them that seriously tainted their relationship. It is possible that Maskelyne's affection for Banks was never

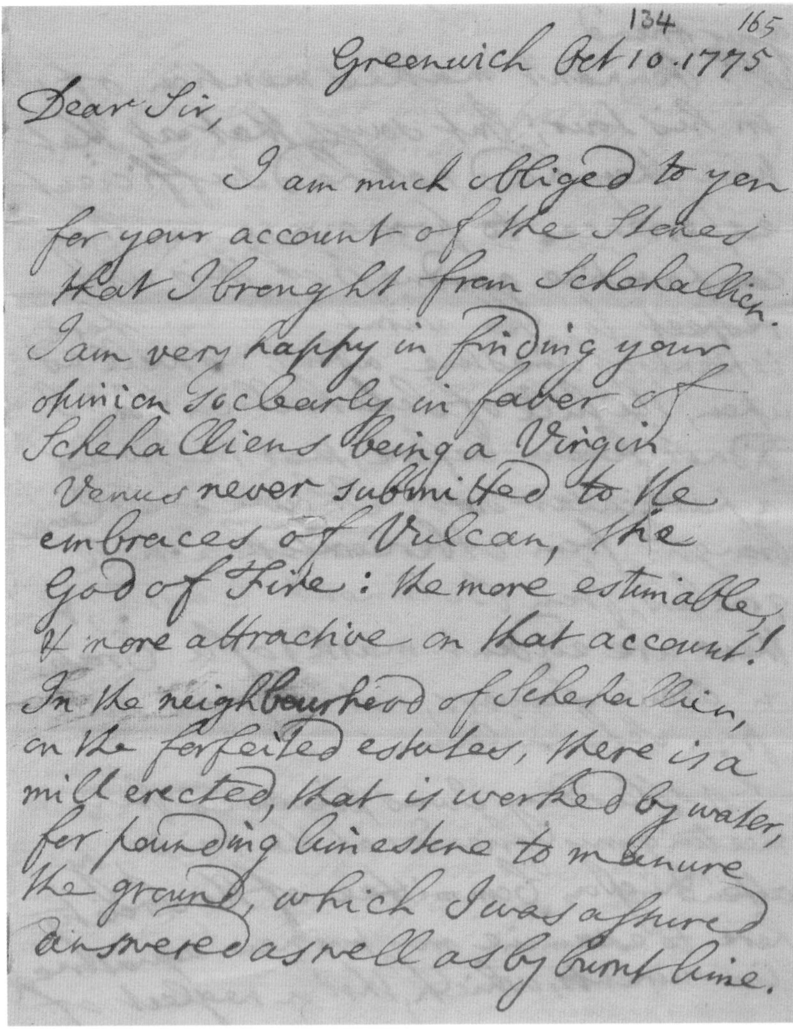

134 165

Greenwich Oct 10. 1775

Dear Sir,

I am much obliged to you for your account of the Stones that I brought from Schehallien. I am very happy in finding your opinion so clearly in favor of Schehalliens being a Virgin Venus never submitted to the embraces of Vulcan, the God of Fire: the more estimable, & more attractive on that account! In the neighbourhood of Schehallien, on the forfeited estates, there is a mill erected, that is worked by water, for pounding limestone to manure the ground, which I was assured answered as well as by burnt lime.

Fig. 3: Letter from Maskelyne to Banks, 10 October 1775.

fully reciprocated, Banks having little time for religion and the clergy. Maskelyne was also a more conservative man than Banks in many aspects of his life and, while completely affable to most, may sometimes have appeared boring. The first general signs of possible friction show up in the social life of the Royal Society, just a month after this letter. Banks, fed up with the tedious rules of the established members of the Mitre Club, made a break with them and formed a separate dining club with seven other friends.

The minutes of the inaugural meeting, in Banks's handwriting, reveal that a new rule preventing guests from attending on consecutive Thursdays had been the impetus for the breakaway group:

> finding themselves more comfortably served and under fewer instructions than they had usually been at the Mitre Club; ... thinking also that a law newly made in that club, viz. 'that no stranger be admitted on two successive Thursdays' might be attended with disagreeable consequences to such of them as had things of more importance to recollect, than where they had dined the Thursday before.[6]

Three of the others were close friends of his who had often attended the official dining club at the Mitre as his guests, including Captain Constantine Phipps (Lord Mulgrave), with whom he had travelled to Labrador and Newfoundland, Colonel William Roy, the Surveyor-General, and Dr Charles Blagden, who became one of his most intimate friends, being his confidant and greatest ally during the Dissensions.

Their initial resolutions reveal the less-than-gracious spirit of their venture:

> Resolved ...
>
> THAT each member be allowed to bring such of his friends as he shall think worthy to visit the club: a certain Mr Poore having declared this morn to one of the members his disinclination to be concerned in (what he pleased to call) so REBELLIOUS a society as ours, resolved unanimously
>
> THAT the said Mr Poore be incapacitated from ever becoming a member thereof and that no ballot in his favour do entitle him to that honour unless this law be unanimously repealed at a meeting at which every member of the society is present.[7]

The social antagonism generated by setting up their own dining club is hard to measure, though Edward Poore's reaction and subsequent permanent exclusion gives an indication of the disapproval it aroused. The awkwardness was ever-present, as Banks's group met at the same time and place as the Royal Philosophers but in a separate private room. Both groups would then go on to the Royal Society meeting. Both clubs met alongside each other over the next five years, but Banks's presence was the key attraction as a source of interesting conversation and international guests. Without it, the Philosopher's Club suffered lower attendances from members and entertained fewer guests during this period.

It is fascinating to see this aspect of Banks's character acting in such unbridled fashion, clearly revealing his frustration with the decisions of others that got in the way of his wishes. When his new club had been running for eighteen months, he recorded 'That the Club did me, Jos Banks, the honour of ceremoniously voting me Perpetual Dictator', a title he seems delighted to have accepted.[8] One of his biographers branded Banks 'the autocrat of the philosophers', and this major flaw in his youthful character of being 'very used to having his own way' has been recognized as being responsible for ending his involvement with Cook's second voyage, as well as for the actions that precipitated the Dissensions.[9] Having been singled out, it is not surprising that Poore's name appears nine years later during the Dissensions as one of the Fellows willing to speak out publicly against Banks.

It is unknown how Maskelyne viewed the splinter group but he was never a guest in their private room. One of the differences between the groups was financial: Banks's paid a higher rate for their room and probably received better service. Always careful with his money and loyal to his established friends, it is unlikely that Maskelyne would have felt comfortable in the 'rebellious group', should he have been lucky enough to be invited.

It is a strong sign of Banks's charisma that, even while he maintained his independence from the key traditional social group within the Royal Society, the Fellows still wanted him as their President when the position became vacant in 1778. Banks's close friend and fellow traveller in the *Endeavour*, Daniel

Solander, was Treasurer of the Philosopher's Club and always attended, reporting news and gossip to Banks of the high levels of support for him to replace Sir John Pringle. Banks remained a compelling leader whose presence was desired by many, his close friend Charles Blagden writing of one dinner that, 'We had a pretty full club to day, but I cannot say it was very entertaining. Your presence is wanting to animate them.'[10] On becoming President he maintained the separate dining group for several years, finally recombining the two groups in early 1781 owing to dwindling numbers in his own party. The offending rule barring guests from attending on consecutive Thursdays was rescinded.

BANKS AS PRESIDENT

When Banks became President of the Royal Society, a major shift in his relative status with Maskelyne took place. For ten years Maskelyne would have enjoyed a sense of being the older and wiser man, with the authority of the Astronomer Royal if not of the same social class as Banks. Suddenly, not only did Banks take the highest-ranking position in the Royal Society, he also became the active head of the organization to which Maskelyne was accountable as Astronomer Royal; the person he had to call on for funds for instruments and to whom he had to deliver copies of his observations. Banks also became an *ex officio* member of the Board of Longitude, and the two of them were brought even closer in the work they had to do together.

There are subtle signs from the early days of their working relationship that Banks had a poor opinion of Maskelyne's professional performance. When the amateur astronomer, William Herschel, discovered a possible new comet or star in 1781, Maskelyne was closely involved in the process of checking Herschel's results and identifying the nature of the new object, which turned out to be the first planet to be discovered since the Ancient Greeks, now known as Uranus. Mild criticism can be detected in Banks's report to his friend Blagden, as he described the meeting when Herschel read his paper to the Royal Society,

> written for the purpose dedicating his new planet to the King under the name of Georgium Sidus & acknowledging his Majesties Bounty to him. Tho neither the name is well conceivd or the paper well written, it was tolerably well receivd. *The Astronomer* [Royal] *after the meeting got up most unexpectedly to declare his approbation of the name* & thank M^r. Herschell for the benefit he had receivd by his Labors. [emphasis added] [11]

In the months leading up to the Dissensions, Banks made direct criticisms to Blagden regarding Maskelyne's suspected over-promotion of papers published by French mathematicians:

> It is not the first instance of awkward Conduct of the Astronomer royal towards the publications of the R.S. that I have met with. I shall Most assuredly touch him up when we meet by the plain Question, 'did you, M^r A, Know that it had been publishd before?' & if he says yes, 'why did you not tell me so then, as in that case I would not have given my consent to republish M^r. Delalandes works in any shape whatever.'[12]

Throughout the Banks and Blagden correspondence during the summer of 1783 are signs of their mutual wish for Blagden to replace Paul Maty, who was under-librarian at the British Museum, as Secretary of the Royal Society, and hints that Banks was planning to bring this about when meetings reconvened in November.[13] However, Banks's first step in restructuring the secretarial roles in the Royal Society was to force the resignation of Charles Hutton from his post as Foreign Secretary by restricting the post-holder to being resident in London. Hutton was the Professor of Mathematics at the Royal Military Academy, Woolwich, and the man who had analysed Maskelyne's observations from Schiehallion to calculate a value of the density of the Earth. Banks put this motion forward during the Council meeting on 20 November 1783, and Maskelyne and Paul Maty were the only Council members to speak in Hutton's defence and vote against the motion.

The following week, at the anniversary meeting of the full

Society, Maskelyne found he had been omitted from the Council for the following year, without being replaced by an astronomer of similar stature. These two actions were perceived as direct attacks on mathematical Fellows, and provoked a sharp response from those who felt the proud traditions of the Royal Society, as embodied by its former President, Sir Isaac Newton, were in grave danger.

Led strongly by the Reverend Samuel Horsley, a former Secretary of the Society and publisher of Newton's work, they made early gains prior to Christmas, forcing a motion to propose thanks to Hutton (carried by thirty votes to twenty-five), and secondly passing a motion that Hutton had not neglected his duties (forty-five votes to fifteen). Seven of Hutton's supporters made impassioned speeches in his support, including Maskelyne. In Banks's personal notes of the proceedings, he records Horsley saying 'If we cannot do this, we will secede and leave that <u>Bauble</u> [the mace] upon the table, the empty shadow of what the Royal Society was' with Maskelyne's reply, 'Yes, sir, for where the Learning is, there will the Real Royal Society be.'[14]

However, the acrimonious discord caused by these speeches led ultimately to the Huttonites' downfall. Both sides regrouped over the Christmas recess and key figures such as the chemist, Henry Cavendish, who had supported Hutton initially, came round to support Banks's continued presidency, as they wished for calm to be restored to the Society's meetings. After serious canvassing on Banks's behalf by Blagden and invitations by Banks for as many Fellows to attend the meetings in the New Year as possible, a vote of approbation was proposed in favour of Banks's presidency. The number of Fellows attending more than doubled in response to Banks's appeal, and the vote of support for the President was carried by 119 votes to forty-two in early January.

Despite the clear majority support for Banks that this demonstrated, the disruptions continued as the mathematical Fellows attempted to have the 'London-resident' restriction on the post of Foreign Secretary lifted and Hutton reinstated. Their efforts were to no avail: in the end, Maty resigned his position as Secretary and thus Banks's wishes were fulfilled when Blagden was appointed to the post after a vote in early May. Hutton stood against him but lost with only thirty-nine votes to Blagden's 139. The Society had voted strongly for peace under Banks's leadership.

We can only guess at what point Maskelyne accepted that opposing Banks was an untenable position. He continued to campaign against Banks beyond the 8 January vote, signing motions to have Hutton reinstated and to restrain Banks's influence on the elections of Fellows, both of which were defeated in February. He did not vote to support Paul Maty's final stand on 25 March but it is likely that he voted for Hutton rather than Blagden to be appointed Secretary in May. Ultimately, to hold out against Banks when the consequences were clearly futile was neither in his best interests nor in his character.

He was never discussed as a ringleader in the Banks and Blagden correspondence in the same way as Horsley and Maty. Indeed, the consequences for these agitators were effective exile from the Royal Society and penury for Maty, who was never able to recover the income lost with his position as Secretary. However, Maskelyne was not allowed to go unpunished. In an unusually vindictive moment, at a Board of Longitude meeting on 6 March, Banks took the opportunity to castigate the Astronomer Royal in front of six other members for paying bills to stationers and printers that he had no authority for doing, and also threatened to stop working with him.[15] Banks wrote to Blagden with some satisfaction, describing the moment 'when I was revenging myself on the R Astronomer at the board of Longitude who was reducd to a compleat state of humiliation' (Fig. 4).[16]

Resigning from his post as Astronomer Royal to avoid this conflict was never an option for Maskelyne, so he and Banks – having publicly expressed their dislike, distrust and distaste for each other – were forced to begin the task of rebuilding their working relationship. Correspondence from August that year reveals the starched formality that had replaced the more comfortable familiarity of their earlier letters. Banks wrote to Maskelyne informing him that official permission had been given allowing Maskelyne to keep any extra copies of the *Greenwich Observations* printed by the Royal Society as a reward for his

Told that Mr Maty had not signified an
intention yet of resignation

A shocking event has taken place on the
Coast of the Jersey Sr Chas Douglas having been
Teazd by the desertion of several men sent
his Long boat with empty water Casks to
be exchangd for full ones on board a transport
astern the men hauld the boat ahead cut the
Painter & pulld for the shore The first Lieut
Haliburton lady Mortons second son the Lieut of
Marines 11. midshipmen & a common seaman leapd
into the barge & pursue a snow storm came on
they made the Land in the night & attempting
to penetrate inland for shelter were every man
frozen to death the dispatch came from Sr Ch D.
today where I was revenging myself on the ast
astronomer at the board of Longitude & have reducd
to a compleat state of humiliation
Your Faithfully
Jos: Banks

Fig. 4: A letter, dated 6 March 1784, from Joseph Banks to his close friend Charles Blagden, referring to his 'revenging himself' on Maskelyne.

work in supervising their publication. In stark contrast to the open, confident and friendly tone of the 1775 letter regarding Schiehallion, Maskelyne's reply to this information is painfully polite and full of awkward expression. In August 1784 he wrote:

> Sir, I return you my thanks for the intelligence you have favored me with of his Majesty's pleasure relative to the disposal of the

> copies of the Greenwich observations in my favor, after the reservation of such copies as yourself as President with the Council of the Royal Society should chuse for presents; as also for the application you have been pleased to make to his Majesty thro' the Secretary of State which has produced this order.[17]

He then discussed the details of the new arrangements and finished with thanks for the greetings sent to him and his new wife, sending greetings in return to Banks's wife and sister. His expression was again clumsy, perhaps owing to the novelty of writing on behalf of himself and his wife, but it is more likely a result of the awkwardness of having to restore a vital working relationship after a major betrayal. More pointedly, all of Maskelyne's previous letters had begun 'Dear Sir', and finished with the standard signature: 'I remain, Dear Sir, Your most Obedient Servant.' This letter was the first to omit the word 'Dear' from both phrases, thus removing affection from their correspondence.

Notwithstanding the incident in the Board of Longitude meeting, in public Banks was largely conciliatory after the Dissensions. He spoke at the anniversary meeting in November 1784, calling all present 'to throw a veil of oblivion over all past animosities and unite once more in sincere efforts towards the advancement of the Society, the honour and reputation of which we have all equally pledged ourselves to support'.[18] Maskelyne, however, remained excluded from the Council for a second year, a highly unusual slight to the office of Astronomer Royal. This may explain the continuing unease between them suggested by Maskelyne's choice to employ the more formal third person in his correspondence with Banks during 1785. This was a fairly standard writing format between Fellows when discussing papers for the *Philosophical Transactions* but it was also a way to keep a professional distance between correspondents, despite the stilted expression it sometimes produced:

> Dr. Maskelyne presents his complements to Sir Joseph Banks, and should be much obliged to him if he would be so good as to favor the Revd. Malachy Hitchins (a very deserving servant

> of the Board of Longitude, as comparer of the nautical almanac) with his interest with some British peer for a chaplainship.[19]

Blagden continued to monitor the situation and passed on small items of gossip to Banks from his talks with Maskelyne, recording that the 'Mathematicians' were meeting separately from the Royal Society and considering publishing their papers in secessionist transactions edited by Hutton.[20] Occasional signs of their mutual irritation with Maskelyne appear in their letters, including a description of him as seeming 'very sulky'.[21] Then, in a sign that some form of *rapprochement* had occurred, Maskelyne was reappointed to the Council in November 1785. The following year the term 'Dear' also reappeared in Maskelyne's letters to Banks, both at the start and finish of his letters. Some of these dealt with sending the astronomer, William Dawes, to Botany Bay with the First Fleet. The colonization of 'New Holland' (now Australia) was a project strongly supported by Banks, and the venture may have reminded Maskelyne of their early connection in relation to Cook's first voyage. The letter below is perhaps the closest he came to being confident of expressing warm friendship to Banks once again in his correspondence:

> Dear Sir, The Bearer, M^{r}. Dawes, is the person whom I mentioned to you by a letter as desirous to make any observations that might be thought useful at Botany Bay, & whom you approved of in the answer you favored me with.... I at present make all the observations myself, & therefore cannot attend the first meeting of the Royal Society. I must take some opportunity to wait on you of a morning. I am,
>
> Dear Sir,
> Your most Obedient Servant,
> N. Maskelyne[22]

However, in mid-1787, 'Dear' disappeared once more, this time permanently. The change is quite marked, from one letter on 26 May to the next on 26 June. Thereafter, Maskelyne addressed Banks as 'Sir', and any addition to the 'Your most Obedient

Servant' format expressed respect and deference rather than affection. The abruptness of the change indicates the effect of some kind of exchange between them, though what that might have been is not easily apparent. Maskelyne continued to use the third-person format occasionally, and while this does not reflect animosity by any means, it is certainly one of the strongest pieces of evidence that, however much they were able to tolerate each other after the Dissensions, they were never close friends. Charles Blagden was undeniably a member of Banks's inner circle and the only time he used the third-person form of address was during a falling out between them in 1789. Banks was shocked that he did so, and wrote on the letter: 'This is the First time the D^{r}. Ever adressd me thus & after an unbroken Friendship of many years I think it Extraordinary.'[23]

A further sign that social interaction had become more peaceful in the wake of the Dissensions occurred during the final stages of the Paris-Greenwich triangulation project undertaken by (now) General William Roy in August 1787. This Anglo-French project, which aimed to use geodetic methods to resolve any doubts regarding the relative latitudes and longitudes of the Paris and Greenwich observatories, had been suggested by Cassini de Thury in October 1783, just prior to the Dissensions. The task of supervising the project had gone to Roy, a highly experienced surveyor who was also a close friend of Banks and a founding member of the 'rebellious' dining club. Maskelyne was not present during the baseline measurement in the summer of 1784 which was a scientific event of great celebrity, including a visit from George III and Queen Charlotte. Indeed, Maskelyne had very little involvement with the project other than being requested, by virtue of his office as Astronomer Royal, to prepare a response to Cassini's memoir that accompanied it when it was published in the *Philosophical Transactions*.

Delays in completing this inspired a terse letter demanding an explanation from Roy in December 1786, threatening to publish Maskelyne's reply to prove that the non-appearance of Cassini's memoir was not due to himself.[24] Roy believed that Banks had sent the memoir to Maskelyne in early 1784 but the Astronomer

Royal's records suggest he did not receive it until April 1785. His response was duly published alongside the memoir in February 1787. However, despite these tense undercurrents, when the final stages of the triangulation project were conducted at Greenwich in August that year, Maskelyne invited both Roy and Charles Blagden to dine with him. As Blagden wrote to Banks, 'General Roy is now got to Greenwich, & last night I received an invitation from the Astronomer Royal to attend the use of the Instrument there to day, & dine with him; which I shall accept.'[25]

There are fewer letters of Banks to Maskelyne, to explore their patterns of address, but it does seem that his attitude towards the Astronomer Royal mellowed during the 1790s. Two letters written early in the decade began 'Revd Sir', while after 1797 Banks always used 'My dear Sir'. Maskelyne only featured in neutral terms in letters between Banks and Blagden from this time on, and then not very frequently other than in his official role as Astronomer Royal. It was certainly a period when they saw eye-to-eye more often than in any other.

DÉTENTE: THE 1790S

The first of two key events that brought Banks and Maskelyne together was the petition to Parliament brought by Thomas Mudge junior, on behalf of his father Thomas senior, regarding the decision of the Board of Longitude to refuse him a reward for his improvements in the design of timekeepers. In Mudge junior's protest he viciously attacked Maskelyne, claiming he had deliberately misused the chronometers and applied erroneous methods for testing them. Both this and the involvement of Parliament, which Banks viewed as a slur on the professionalism of the Board of Longitude, provoked sharp responses from Banks in defence of Maskelyne as well as the Board. He wrote to Mudge, 'I take this occasion however to add that having read your pamphlets and the Ast^{r} Royals Answer, I am decidedly of Opinion that your attack upon his character was highly unjustifiable.'[26]

When Maskelyne was next required to test the performance of Mudge's timekeepers on behalf of the Board, he requested to use the same conditions as for Harrison's H4, which required witnesses to the process. Banks wrote to him supportively:

> Revnd Sir,
> I cannot wonder that after the unprecedented attack made upon your Character by Mess Mudge & Graham you are desirous to shelter yourself in Future behind the entrenchment of regular & Cautious measures rather than give way to that spirit of liberality which inducd you of late to try the watches Sent by the board to the Observatory with as little Trouble to their maker and as little expence to the Public as Possible.[27]

Despite their combined efforts, and Banks's masterful diplomacy when appealing to the Parliamentary committee to respect the decisions of the Board, Mudge won his petition and was awarded £3,000 in prize money. The experience of having a common enemy and fighting for the same cause certainly brought Banks and Maskelyne closer.

Their second, more successful, project was the matter of publishing the observations of the third Astronomer Royal, James Bradley, a goal that Maskelyne had set for the Royal Society following Bradley's death in 1762. It had hitherto been the habit of the Astronomers Royal at Greenwich to consider their observations their own property, the unfortunate consequence being that the documentation was kept by their families rather than being published and made available to other astronomers and mathematicians. This long process of making Bradley's observations publicly available, which involved possible court action in the 1770s, came to a stop when the papers were presented to Oxford University for publication under the editorship of Thomas Hornsby, Savilian Professor of Astronomy and a Commissioner of Longitude.

In the mid-1780s Banks, as President of the Royal Society and on behalf of the Board of Longitude, made enquiries into the causes of the apparent delay in publishing the material. Despite putting some pressure on Dr Hornsby, no explanation more specific than 'ill health' was forthcoming and no alternative avenues of getting the observations published were found. Stronger action was taken in the 1790s, when the Board of Longitude printed a pamphlet describing the efforts they had to date made on the matter, which was circulated among 'enemy camps' to shame them into action. Maskelyne's excitement at this was palpable – he had so far been trying to get access to Bradley's observations for thirty years. He wrote enthusiastically to Banks to ask advice and discuss strategies for delivering the pamphlets to achieve maximum effect.[28]

But Maskelyne's highest praise for Banks throughout their working relationship was expressed with delight when the first volume of the observations was finally published. It included an indirect reference to one of the fears expressed during the Dissensions, that under Banks's leadership mathematics would be passed over in favour of botany:

> Sir, ... I have this instant received a letter from M$^{r.}$ Robertson, D$^{r.}$ Hornsby's Colleague, that the first Volume of D$^{r.}$ Bradley's observations is finished; ... I congratulate you on this pleasing intelligence, and doubt not you will think as I do that the spirited and well directed exertions of the Board of Longitude and Royal Society ... have been the main cause of this publication.... As a person principally interested in this publication by my situation here [at Greenwich], I shall ever acknowledge your kind and public spirited assistance in this business, worthy of the President of the Royal Society, *and his accustomed zeal to promote every branch of Science.* I am,
>
> Sir,
> *with much respect*, [emphases added]
> Your most Obedient Servant,
> Nevil Maskelyne [Fig. 5].[29]

It would seem that after twenty years they had found a peaceful way of working together productively. Unfortunately, despite the high levels of co-operation, support and mutual respect

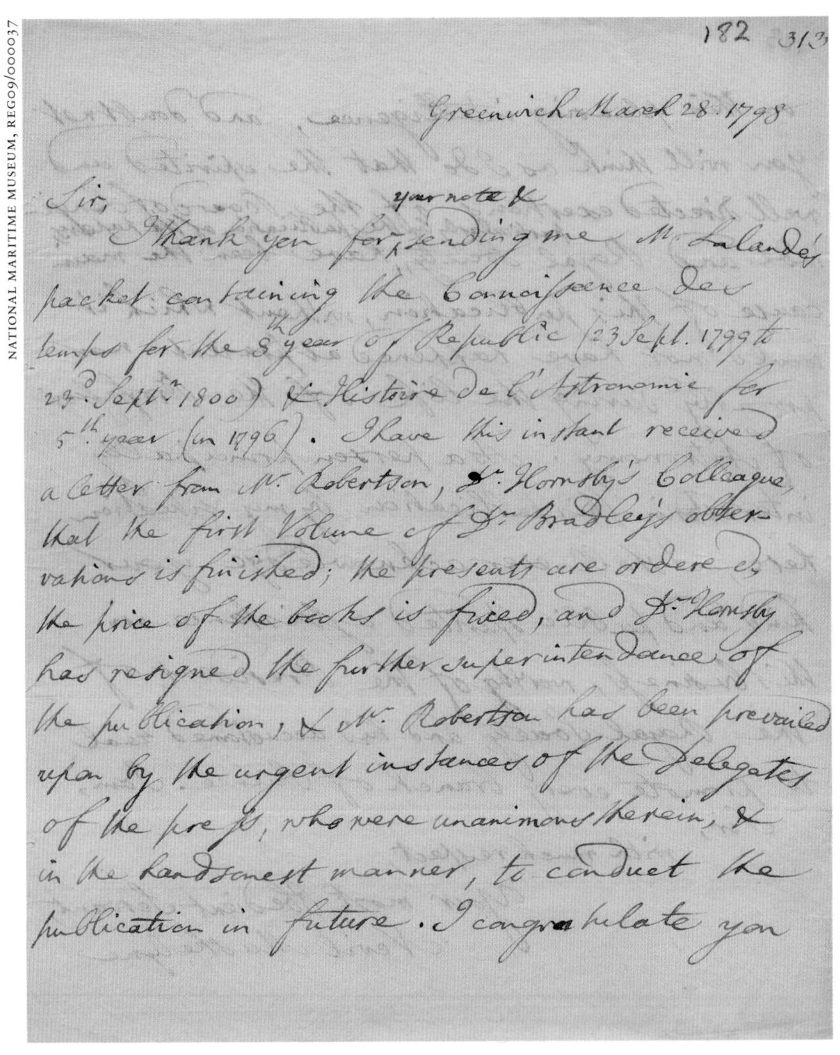

182 313

Greenwich March 28. 1798

Sir,

I thank you for your note & sending me Mr Lalande's
packet containing the Connoissance des
temps for the 8th year of Republic (23 Sept. 1799 to
23d Septr 1800) & Histoire de l'Astronomie for
5th year (in 1796). I have this instant received
a letter from Mr Robertson, Dr Hornsby's Colleague,
that the first Volume of Dr Bradley's observations is finished; the presents are ordered,
the price of the books is fixed, and Dr Hornsby
has resigned the further superintendance of
the publication, & Mr Robertson has been prevailed
upon by the urgent instances of the Delegates
of the press, who were unanimous therein, &
in the handsomest manner, to conduct the
publication in future. I congratulate you

Fig. 5: A formal but complimentary letter from Maskelyne to Joseph Banks, dated 28 March 1798.

demonstrated in these two events, there was one further major conflict for them to negotiate that had serious ramifications for the Board of Longitude.

183

on this pleasing intelligence, and doubt not
you will think as I do that the spirited and
well directed exertions of the Board of Longi-
particularly by the publication of the papers,
tude and Royal Society, have been the main
cause of this publication, without which it
would not have happened at present, nor
probably during the life of the Professor
of Astronomy. As a person principally
interested in this publication by my situation
here, I shall ever acknowledge your
kind and public spirited assistance in
this business, worthy of the President of
the Royal Society and his accustomed zeal
to promote every branch of Science. I am,
Sir,
with much respect,
Your most Obedient Servant,
Nevil Maskelyne

NATIONAL MARITIME MUSEUM, REG09/000037

EARNSHAW V ARNOLD, OR MASKELYNE V BANKS

The highly charged priority dispute between watchmakers Thomas Earnshaw and John Arnold, discussed in the previous chapter, had significant consequences for Maskelyne and Banks. For many years Banks had been a supporter of Arnold, a watchmaker who began to produce large numbers of affordable marine timekeepers through the significant improvements he made on Harrison's original design. Then Maskelyne reported to the

Board of Longitude in March 1803 on the results of recent tests and the Board resolved that, since Earnshaw's timekeepers had 'gone better than any others that have been submitted to trial', he deserved 'a reward equal at least to that given by Parliament to Mr Mudge'. Banks, in consequence, was deeply concerned that Arnold's achievements were being sidelined by an inferior watchmaker.[30]

Banks found the proceedings of the next few years greatly to his distaste. Maskelyne himself was initially concerned with errors in his calculations and the possibility that two of Arnold's watches might actually have performed better, so requested an extraordinary meeting of the Board to verify his judgement of Earnshaw's work. Banks noted that this cost the public £90, but was justified given the money at stake. He was thus greatly surprised when, rather than acknowledging his errors at the second meeting, Maskelyne produced new calculations in support of Earnshaw's watches over Arnold's. The Board then confirmed their previous decision to award Earnshaw the prize. Banks recorded his response to this:

> On reconsidering the Proofs the Astronomer [Royal] had given to the extraordinary Board they appeard to Sir Joseph fallacious at least, if not unfair, he therefore applied to Mr Gilpin [the Board's secretary and a former assistant to Maskelyne] for his opinion, who confirmd the fallacy of them & undertook at Sir Joseph's particular desire to make a comparison between the Watches for Sir Joseph's use.[31]

This event seems to have reignited the annoyance Banks felt towards Maskelyne in several ways. First, there was his tendency to double-check himself after making an initial judgement, causing delays and extra work for those around him. Secondly, while the quality of his work and international reputation suggests that Maskelyne was usually reliably accurate in his calculations, Banks had no tolerance for any errors that he did make, real or otherwise. Banks attended the next meeting of the Board with Gilpin's calculations in hand, determined to convince them of the grave error of judgement they were in danger of making, but to no avail.

A year later, as the dispute between the watchmakers carried on, Banks published Gilpin's calculations in his *Protest against a Vote of the Board of Longitude*, and set himself apart from the rest of the Board in their decision to grant Earnshaw a reward. Maskelyne, in response, published an *Answer to Sir Joseph's Protest*, in which he outlined his calculations and reasons for choosing Earnshaw as the better watchmaker. For the first time since the Dissensions, their disagreements were being aired in public. The Board ultimately voted for a compromise, ruling in December 1804 that both Earnshaw's and Arnold's work was worthy of an award of £3,000 each.

Up until this time Banks had been one of the most reliable members of the Board, attending seventy-two out of the previous eighty meetings. But when the next extraordinary meeting of the Board to consider this matter fell due in July, Banks was unavailable for it owing to 'pressing business'. As a result the Chairman asked for a delay until the December meeting. Banks may have had good reason to absent himself but his 'pressing business' did not keep him from attending the Royal Society's dining club (with Maskelyne) that same day. Banks had also missed the previous meeting on 6 June and was absent from the next on 12 December, when the Board decided to go ahead without him and confirmed their final compromise recommendations. While this last absence was most likely the result of a debilitating attack of gout that kept Banks bedridden for forty days at the end of that year, missing the previous two consecutive meetings suggests that he was deliberately absenting himself from the Board over this issue.

In the following February, Earnshaw published an article in the *Morning Chronicle* full of invective against Banks. A few weeks later Maskelyne attempted to placate the latter, advising him to take no notice of Earnshaw.[32] To this Banks replied:

> My dear Sir ... I conclude from the perusal of the Letter that you have not carefully looked over M$^{r.}$ Earnshaw's advertisement. I think, if you had, you would not have advised me to set quiet

> under a Public accusation, (from a Man whose Talents have entitled him to a large Public Reward,) of being *the head of a Combination of wicked & malicious men conspiring together to oppress an individual.*[33]

Banks sought legal advice and received the professional opinion that Earnshaw's words were libellous. He drafted a resolution to the Board and attended the meeting on 6 March 1806, putting forward a motion that it should take legal action against Earnshaw in defence of his own character. The Board recorded its strong disapproval of Earnshaw's behaviour but declined to take formal legal action. If Banks had any confidence that this group of mathematical clergy would support him this way, he was sorely disappointed, and perhaps the more so by Maskelyne, whom he had vigorously supported ten years earlier during the period of Mudge's attacks on him.

This was the last meeting of the Board that Banks attended for five years. He did not return until the first meeting after Maskelyne died in February 1811. There can be no other interpretation of his absence than his refusal to work any longer with Maskelyne in this context, finally fulfilling the threat he had made in March 1784, during the Dissensions. Banks had always been sorely irritated by the errors that he perceived Maskelyne to have made: finding that the Astronomer Royal and the various professors would not take action against Earnshaw on his behalf, he no longer had the grace to do business with them (Fig. 6).

THE AFTERMATH

While the Earnshaw/Arnold dispute had clearly pushed the limits of Banks's tolerance of Maskelyne as far as Board of Longitude business was concerned, the two men continued to socialize within the Royal Society and, as President, Banks continued to oversee the activity of the Royal Observatory. A letter from Maskelyne in 1808 indicates that some personal connection remained, since he felt free to add a postscript referring to

THE ROYAL SOCIETY, P0005

Fig. 6: A large oil portrait of Joseph Banks by Thomas Phillips, 1808. He is depicted in the Presidential chair of the Royal Society, with the Society's mace before him, and wearing the ribbon and star of the Order of Bath.

Banks's recurring attacks of gout, but his general tone of address continued as one of deference and respect rather than affection.[34] Earnshaw went on to make further petitions for rewards directly to Parliament and, when Maskelyne wrote to inform Banks of the outcome in 1809, he once again took up the third-person mode. He announced that the Board of Longitude had been cleared of any blame and acknowledged that in a recent comparison of Arnold's and Earnshaw's watches, Arnold's had gone best – hoping, perhaps, that this would satisfy Banks's wish for due recognition of Arnold's talents over Earnshaw's.[35]

But even with such a peace offering, Banks revealed his lingering displeasure a year later by arguing less forcefully than he could have done in favour of a pay rise for Maskelyne, who had been receiving the same salary of £300 a year since 1765. Banks reported back to Maskelyne on his approach for an increase to the Board of Ordnance which, while admitting that it would be appropriate, considered that

> it [would be] an easier matter to settle the quantum of increase ... with your Successor than with you, to whom personally it could not be any great object, but he [the Prime Minister] ended by saying, now do not you think Sir Joseph that there are able Men both of Oxford & Cambridge who would be happy to undertake the business of the Royal Observatory for £300 a year & a good House? *to which I did not venture to deny my assent* [emphasis added].[36]

John Pond's appointment as Astronomer Royal a year later was met with a doubling of the salary to £600. It is true that Maskelyne was significantly wealthier at the end of his life than when he started out at Cambridge as a sizar: through careful management of his investments, inheriting the estates of his childless uncles and older brothers, and marrying well, he left an inheritance for his only daughter, Margaret. that was described as giving her 'dazzling expectations'.[37] That said, his resources were modest compared to the landed wealth Banks had enjoyed from the beginning of his life, so this final lack of support from the latter was somewhat churlish.

When Maskelyne died at the age of seventy-nine, Banks wrote with sympathy to his wife and daughter, honouring him and acknowledging the

> loss they have sustained, which great as it is can only be compared with the loss Science has suffered by the Death of one who has done more for the true interest of his Country by introducing & promoting the use of Correct astronomy in [the] navy ... [more] possibly than any person who has hitherto occupied the Elevated Stations in which he was Held.[38]

It is possible that Banks was merely being polite and, as President of the Royal Society, paying appropriate lip-service towards one of its longstanding members; but it is more likely (and more generous to believe) that his words were a genuine expression of the respect he held for Maskelyne's scientific work. Maskelyne and Banks were brought together by their deep interest in learning and their wish to make the pursuit of scientific knowledge the centre of their lives. While Maskelyne enjoyed Banks's company and encouraged his involvement in the Royal Society in the early years, Banks seems to have found Maskelyne irritating, particularly after he became President. Their social, cultural and academic differences surfaced dramatically during the Dissensions of 1783–84, and left an indelible mark on their relationship thereafter.

Nevertheless, as Banks matured he found a greater level of tolerance for the Astronomer Royal and they were in agreement far more often in later years, until the Earnshaw and Arnold dispute brought an unfortunate end to their work on the Board of Longitude. One could wish, for their sakes, that their friendship had been closer and more easily sustained and that working together had been more enjoyable. Instead we can applaud the mutual respect for each other's contributions to science, which sustained both men through the tense disagreements they endured and which enabled them to work together as productively as they did.

FURTHER READING

Richard Holmes provides an engaging introduction to the period and Banks's role within it in *The Age of Wonder: How the Romantic Generation Discovered the Beauty and Terror of Science* (London: Harper Press, 2009). In-depth biographies of Joseph Banks include H.C. Cameron's concise *Sir Joseph Banks: The Autocrat of the Philosophers* (London: Batchworth Press, 1952) and Harold B. Carter's exhaustive *Sir Joseph Banks 1743–1820* (London: British Museum Natural History, 1988), the fruit of twenty-five years of research.

Neil Chambers has undertaken to transcribe and edit Banks's voluminous scientific correspondence, and this chapter would not have been written without this resource: *The Scientific Correspondence of Sir Joseph Banks, 1765–1820* (London: Pickering & Chatto, 2007).

John Gascoigne has written extensively on Banks and the period, two works touching on the issues with Maskelyne and the Royal Observatory while focusing on the wider sphere of influence Banks maintained as President of the Royal Society: *Joseph Banks and the English Enlightenment: Useful Knowledge and Polite Culture* (Cambridge: Cambridge University Press, 1994) and *Science in the Service of Empire: Joseph Banks, the British State and the Uses of Science in the Age of Revolution* (Cambridge: Cambridge University Press, 1998). The Dissentions are discussed fully in J. L. Heilbron, 'A mathematicians' mutiny, with morals', in *World Changes, Thomas Kuhn and the Nature of Science* ed. P. Horwich (Cambridge MA: MIT Press, 1993), pp. 81–129.

For further information about the Dining Clubs of the Royal Society, see T.E. Allibone's *The Royal Society and its Dining Clubs* (Oxford and New York: Pergamon Press, 1976).

CASE STUDY G

VISUALIZING AND COLLECTING THE MASKELYNES

In recent years, the National Maritime Museum has been fortunate enough to have received – by gift, donation in lieu of inheritance tax and purchase – a significant number of objects, books and manuscripts relating to Maskelyne and his family. The manuscripts have been digitized and are openly accessible, and the books are in the Caird Library, where they can be inspected on request by members of the public. The books include Maskelyne's own copies of the *Nautical Almanac* and the *Tables Requisite* but also, engagingly, Le Sage's picaresque *Les aventures de Gil Blas* and Defoe's marooned *Robinson Crusoe*.

The Museum had already possessed many of the instruments used by Maskelyne and his assistants since the 1950s, as a result of having then taken over the historic Royal Observatory buildings at Greenwich where he lived and worked, including much of its early equipment. As well as the suite of transit instrument, clock, quadrant and zenith sector that came to the Observatory in Bradley's time, there are also the clock and mural circle that were ordered by Maskelyne toward the end of his life but which only came into use after John Pond became Astronomer Royal (Fig. 1). This was fitting in some ways, given that it was Pond's calculations which alerted Maskelyne to the errors in his instruments and prompted his move to replace them.

The recently acquired collections include some additional Maskelyne-associated instruments, but these are of a very different kind. Although they too underline the importance of the London instrument trade in this period, they are smaller items that relate to personal use and leisure. Some were purchased by Maskelyne's sister, Lady Clive: a three-inch refracting telescope made by Dollond of London in the early nineteenth century (see Fig. 3, p. 10), with which she observed the planets regularly and

Fig. 1: Mural circle by Edward Troughton, 1810. This was ordered for the Royal Observatory, to replace the two mural quadrants, while Maskelyne was still alive. It was intended to be used to measure both celestial co-ordinates of a body crossing its meridian, although it was ultimately used to establish vertical positions while a transit instrument was used to establish right ascension (east-west position).

attempted to teach her family about astronomy, and a pair of globes (Fig. 2).

The globes were dedicated by the makers, W. & S. Jones, to the Astronomer Royal, and Margaret Clive reported to him that 'I have received the beautiful pair of Globes. Nothing can exceed their loveliness, and Messrs. Jones ought to be told (and pray tell them) how much they are approved by their astronomer of Oakley Park.'[39] Probably also Lady Clive's were three spyglasses, which are of high quality and, in two cases, signed by their makers, Dollond and Dudley Adams (Fig. 3). Two more significant instruments are watches by William Coombe and Thomas Earnshaw, which belonged to Nevil Maskelyne and are discussed in Chapter 5.

Rather more humble in appearance are a compass in a glazed wooden box, and a boxwood quadrant. The former might have been Lady Clive's, or have belonged to another member of the family, but the latter has a manuscript note attached to it that says 'This sextant [*sic*] belonged to Dr. Nevil Maskelyne D.D.' If it did, it was certainly made before he was born and was more likely retained as a curiosity or keepsake rather than as a functional item. Far more decorative was a small planetarium that,

NATIONAL MARITIME MUSEUM, ZBA5106, ZBA5107

Fig. 2: This pair of celestial and terrestrial globes was made by W. & S. Jones, for Lady Clive.

NATIONAL MARITIME MUSEUM, ZBA5109–11

Fig. 3: These late eighteenth-century spyglasses probably belong to Lady Clive. They are fashionable and high-quality items, and were kept by the family until acquired by the National Maritime Museum. The shagreen and horn example is signed by Dollond, the nine-draw nickel-silver by Dudley Adams and the ivory inlaid spyglass is unsigned.

according to family tradition, belonged to Maskelyne's daughter, Margaret (Fig. 4). This was made by W. Jones at some time in the first decades of the nineteenth century, perhaps while Maskelyne was still alive and Margaret resident at the Royal Observatory. It is an example of a fairly common type of instrument, for educational and ornamental use, nicely displayed on a turned wooden stand.

Although, as many chapters in this book amply demonstrate, the manuscripts at Greenwich and Cambridge are very revealing, the items that probably best help to bring the Maskelynes alive in our imaginations are those which help to evoke them physically. There are several portraits, a number of items of clothing – which really do help us to imagine the Observatory as a space in which these individuals were once present – and some drawings by the young Margaret, through which we can vividly imagine her at work in or near her home.

The number and quality of the portraits in this collection testify to the eventual prosperity of the family and to their social

NATIONAL MARITIME MUSEUM, ZBA4664

Fig. 4: Made at the very end of the eighteenth century, or in the early nineteenth century by W. Jones, this planetarium orrery was kept by the family and is said to have belonged to Margaret Maskelyne. It is typical of Jones's portable orreries, but provided with an elegant turned stand.

connections. Three are by John Russell, who was a pioneer and leader of English pastel portraiture, and the author of *Elements of Painting with Crayons* (1772). He exhibited at the Royal Academy from 1769, was elected an Associate in 1772 and a full Royal Academician in 1788. The earliest is a head study of Maskelyne (see Fig. 6, p. 20), drawn about 1776 in black, red and blue chalks with great immediacy and freshness, and with the face far more worked up than the wig or clothing. Russell – himself a notable amateur astronomer – was also to portray Maskelyne much later in life (1804), this time in a pair of portraits with his wife (Fig. 5 and see Fig. 7, p. 291).

NATIONAL MARITIME MUSEUM, ZBA5100

Fig. 5: Pastel portrait of Nevil Maskelyne by John Russell, 1804.

Then aged seventy-two, Maskelyne appears to have lost some of his sparkle and we can, perhaps, see why Lady Clive reported to her niece that 'Nobody is pleased with your father's little portrait.'[2]

As in his large, formal, oil portrait by Louis van der Puyl, discussed and reproduced in the opening pages of this book, Maskelyne is shown with classical pillars and drapery, and a view of the Royal Observatory behind. His hand rests on a book, which must surely again be his published *Observations*. The portrait of Sophia Maskelyne feels rather livelier and, as Amy Miller describes in the next chapter, shows her as very much the fashionable matron. Russell also painted a pastel portrait of William Herschel under similar circumstances, and a number of other men of science, including Joseph Banks, but his output was prodigious and he largely catered to the wealthy middle classes. At the height of his fame he could charge thirty guineas for 'a head', making his gift to Maskelyne a generous one.

There are two portraits associated with William Owen, who had studied at the Royal Academy schools and been influenced by the aesthetic theories of Richard Payne Knight and portraits by Thomas Gainsborough and Joshua Reynolds. They are both of the Margarets; Maskelyne's sister and daughter. The first is an 1811 copy taken from a larger portrait of the former by Nathaniel Dance, painted in the 1770s, although Owen had produced his own portrait of Lady Clive some years earlier. She herself had said that Dance's original, painted not long after her husband's death in 1774, showed her 'poor eyes that bore the traces of grief' and her 'face worn by sorrow', yet it was fine enough to merit copying.[3] Owen's original portrait of Maskelyne's young daughter Margaret was painted in about 1793 (see Fig 6, p. 290). In a convention common for portraits of young girls, the figure is placed in a landscape accompanied by a dog and – like Maskelyne himself in his portraits by van der Puyl and Russell – she is also shown in front of her home, the Observatory, as her father's heir but to a property she could not keep.

Margaret's association with the Observatory is also clear

from her drawings of, and from, the buildings and within the Royal Park at Greenwich. A drawing of Flamsteed House produced at the age of sixteen shows the influence of the drawing lessons she received, making this Stuart oddity look like a Palladian country house. She also drew, or began to draw, a panoramic view north from Greenwich Park, probably from just outside the Observatory. The precision of the outlines, and the unfinished nature of the intended shading and colouring, suggest that this was produced with the assistance of a camera lucida, a portable optical device that, with practice, the artist could use to view a superimposed image on the drawing surface (Fig. 6).

Drawings of and from Greenwich Park are a long tradition, and the view was one of which Maskelyne took advantage with the permanent camera obscura that he installed in the western turret of Flamsteed House. Some engraved prints were produced by professional artists using this device.

There is no record of who constructed Maskelyne's camera obscura, but we may remember his early encounter with that belonging to James Ayscough, which enabled him to watch the

Fig. 6: Unfinished drawing of the view from Greenwich Park, by Margaret Maskelyne. The presence of the colonnades around Queen's House shows that the drawing dates to after 1812, by which time the buildings belonged to the Royal Naval Asylum. The Royal Greenwich Hospital is behind.

1748 solar eclipse. Ayscough, like many other optical instrument makers, sold 'Camera Obscura's for delineating landskips and prospects'. David Kinnebrook, one of Maskelyne's assistants, described the set-up at Greenwich:

> at the western corner [of the Great Room] is a Camera obscura, upon a very curious construction, the heighth [*sic*] of the turret upon which the machinery is fixed is about 15 feet, and is capable of containing 8 people; but 6 can stand in it very conveniently, on the top of the turret is placed a cubical box the side of which is about half a yard, in this box is a mirror about 6 inches by 4 which reflects the objects into a lens of about 3 inches in diam^{r}., this lens is fixed in the bottom of the box; and the reflection of the objects pass through this lens unto a concave circular piece of plaister of Paris which is fixed into a circular frame at about 4 feet from the bottom of the turret, but this frame may be elevated or depressed according to the distance of the objects that you want to observe.

He added that, 'by this Camera you may sweep the whole horizon, and as the Camera from its situation commands a most beautiful prospect, it is really a delightful sight on a fine day'.[4] Kinnebrook also told his father that the Great Room, as well as containing three telescopes, displayed the van der Puyl portraits of Nevil and Sophia Maskelyne, a bust of Isaac Newton and several good prints.[5] The inventory of Flamsteed House, drawn up after Maskelyne's death, also tells us that there were other art works in the living quarters of the building, especially in Margaret's room. Some of these may have been her own framed drawings; others purchased.

There is one drawing in the National Maritime Museum collection which may be a portrait by Margaret of her father, although it was probably done posthumously (Fig. 7).[6] It is a design for a cameo and shows a classical profile and a crescent

Fig. 7: This drawing, which appears to be a design for a cameo, commemorates Nevil Maskelyne. Found among the family papers among other drawings and correspondence, it was probably drawn by Margaret Maskelyne.

NATIONAL MARITIME MUSEUM, REG09/000037

moon. The inscription reads *Mare Praestat Eunti*, which is from Ovid: *Venus, orta mari, mare praestat eunti.* Alexander Pope translated this as 'Venus for thee shall smooth her native main': a more recent rendering is, 'Venus, born from the sea, smoothes the waves for a lover'. As a motto for Maskelyne it may be read as 'He smoothes the waves for them' or perhaps even 'they [Maskelyne and/or the Moon] guide them across oceans'.[7]

The following chapter looks in greater detail at the home life of the Maskelynes and, in particular, pulls together the little we know about Sophia. Some of this comes from her portraits and surviving items of clothing, as discussed there. As well as the textiles associated with Sophia, the NMM collection includes two dresses and a fashionable yellow silk spencer, or short jacket, belonging to Margaret (Fig. 8), and also her banyan, or dressing gown, made of striped, padded Indian silk. This fabric is similar to, but not exactly matching, that used in another, unique item in the collection. This is a padded jacket of about 1765 with trousers that have integrated feet, somewhat like a giant 'babygro', and has been designated Maskelyne's 'observing suit' (Fig. 9).

NATIONAL MARITIME MUSEUM, ZBA4682, ZBA4684

Fig. 8: This sprigged muslin dress and yellow silk spencer, kept by the family, undoubtedly belonged to Margaret Maskelyne. They are well made and fashionable items of the early nineteenth century, with the spencer in Charles I revival style.

NATIONAL MARITIME MUSEUM, ZBA4675–6

Fig. 9: Maskelyne's unique padded 'observing suit', dated to around 1765, is made from striped yellow and red Indian silk. While it has been suggested that either the silk or the suit were sent to him by his sister and brother-in-law, Robert Clive, from India, this kind of imported fabric was also readily available in England.

It is certainly well worn and the amount of wear and repair to the very ample seat may be indicative of its use on an observing chair, with the astronomer shifting and then maintaining his position to await the transit of a particular star. The wear on the feet suggests that Maskelyne wore pattens, or raised wooden overshoes, as he walked from house to observing rooms. Although on cold nights the whole ensemble would have been covered with a greatcoat, he must have sometimes cut an amusing figure in his Greenwich home.

8

THE MASKELYNES AT HOME

Amy Miller

LIFE IN THE ROYAL OBSERVATORY, GREENWICH

When Nevil Maskelyne was appointed fifth Astronomer Royal in 1765, he took up residence at the Royal Observatory, Greenwich, in the living quarters that were a perquisite of the post, now known as the Flamsteed House apartments. Nearly twenty years later, on 21 August 1784, he married Sophia Pate Rose of Cotterstock, Northamptonshire, at St Andrew's, Holborn. In the following year their daughter and only child, Margaret, was born at the Royal Observatory. What is apparent in Maskelyne's journals and memoranda from his marriage to his death was that the Royal Observatory had become a family home (Fig. 1).

These journals are fascinating documents for the late eighteenth and early nineteenth centuries in that they are expansive records of the minutiae of day-to-day life: problems with smoking chimneys, food stocks in the larder, where he preferred to buy his shirts and, of course, Margaret's educational progress. They provide insight into the inner workings of domestic life in the Royal Observatory, of a middle-class family, and to a certain extent reflect the way in which use of the Observatory's public and private space overlapped. They also present an interesting juxtaposition of comments on the Board of Longitude, or the merits of various timekeepers, interwoven with recipes for Norfolk

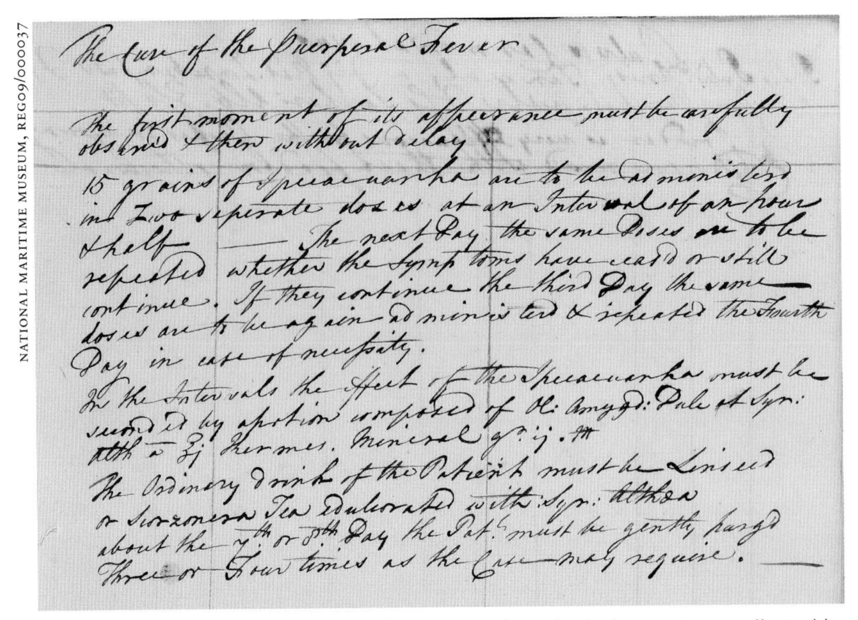
The Cure of the Puerperal Fever

The first moment of its appearance must be carefully observed & then without delay
15 grains of Ipecacuanha are to be administered in Two seperate doses at an Interval of an hour & half — The next Day the same Doses are to be repeated whether the Symptoms have ceased or still continue. If they continue the third Day the same doses are to be again administered & repeated the Fourth Day in case of necessity.
In the Intervals the effect of the Ipecacuanha must be seconded by a potion composed of Ol: Amygd: Dulc et Syr: Altth a ℥j Kermes. Mineral gr: ij. M
The Ordinary drink of the Patient must be Linseed or Scorzonera Tea edulcorated with Syr: Althea about the 7th or 8th Day the Pat.t must be gently purged Three or Four times as the Case may require. —

Fig. 1: Notebook containing recipes for cures and medical observations, collected by Maskelyne between 1766 and 1810. This page shows a recipe for the 'Cure of the Puerperal Fever'.

dumplings, burn ointment, cider and the ownership of various plots of land in villages in Wiltshire and Northamptonshire.

An examination of the journals illustrates the way in which family connections were exploited on an almost daily basis. At the Royal Observatory the Maskelynes regularly received shipments of food from family in the country. Nevil's sister, Margaret, who had advantageously married 'Clive of India' in 1753, sent hampers of game and produce from their country estate to the Greenwich household. Sophia Maskelyne's sister, Letitia Booth, who maintained a country property at Cotterstock in Northamptonshire, likewise sent the family local cheese from there. Maskelyne recorded the hampers which arrived from the countryside with surprising regularity: in November 1790, they received a hare and haunch of venison from Lady Clive, half a buck from Cotterstock and on 18 November, a pheasant from Mr Eaglestone. They also enjoyed quite a fashionable diet, as Nevil Maskelyne recorded in his notebook on 7 November

1798 the arrival of '4 lb Souchong, ½ lb Green and 4 lb black tea'. In June 1800, the family ordered 'Coffee 1 lb, Green Tea 1 lb, 3 lb cocoanuts'.[1] The routine domestic arrangements of the Observatory are recorded in Maskelyne's journals with a focus on issues around the upkeep of the building. In particular, smoking chimneys appear to have presented an ongoing challenge for which various 'home remedies' were deployed, such as the following, which he took from the *St. James's Chronicle* in January 1803:

> A cheap easy and safe method of clearing chimneys of soot. Mix 3 parts salt petre, 2 parts salt of tartar and 2 parts flower of brimstone ... over a strong clear fire near the back of the chimney. If you have not a mind to hear the sound of the report, which will be as loud if not louder than that of the discharge of a gun, get away; and as soon as it begins to boil brown, it will cause such an explosion, as by the mere action of the elastic air in the chimney will, without the least damage or danger, bring down the soot as well or better than when swept by hand.[2]

He notes that 'Mr. Ferminger has cleaned the Observatory chimney, this way with entire success' but, although their brickwork was repointed by the Admiralty in 1789, they continued to plague the Maskelynes and are a regular feature in Nevil's journal entries throughout his forty-four-year tenure as Astronomer Royal.[3]

Foremost in the journals is a preoccupation with the relative health and well-being of the inmates of the Royal Observatory. As might be expected from an astronomer, this is broken down into a series of observations, which tend to centre on his daughter Margaret. In July and August 1791, six-year-old Margaret appears to have contracted a high fever and initially the course of treatment was a rhubarb purge. However, as the fever continued over the course of two weeks, the entries follow its progress:

> She has been growing worse every day. Her fever is high in the night, remits & is less in the day, sometimes comes on and goes

> off in a quarter of an hour; her pulse in the fever is very quick; her skin feels moist for the most part. Saturday night August 6 she sweated a good deal for an hour yet was restless all the while. This morning at 8 am her pulse beat 144 in a minute, and very low & languid.[4]

In addition to being treated with 'spirit of cinnamon and menthol', Margaret was also taken for an 'airing' in the carriage lasting over two hours. Although her fever returned, the fresh air appears to have worked and she began to recover. Her series of childhood illnesses are painstakingly plotted through the diary; every treatment noted, how much and at what times it was administered, what she ate and, most importantly, how she began to mend.[5]

Maskelyne himself appeared to have a similar illness in August 1791, when he gives an 'Account of my own health' in which he records the onset of fever and sore throat and his treatments – taking saline draughts and gargling with brandy, water, and sal volatile in 'equal parts'.[6] He regularly suffered from stomach complaints, as he records in November 1789: 'Great pain in the stomach at night at Dr. Herschel's after drinking some Burgundy.'[7] Apart from being a means of tracking the ailment and its cure, this obsession with observing and recording illness and treatments is, in itself, apparently a measure of his concern and affection for his wife and daughter (though he also discussed such topics with Joseph Banks, with whom his relationship was not always smooth, as described in the previous chapter). Surviving correspondence between Maskelyne and his daughter confirms their very close and affectionate relationship, while his wife, Sophia, also figures in closely observed treatments for her health. She is referred to affectionately as 'Mrs. M' while Margaret is known as 'Miss Margaret' or 'Miss Margaret Maskelyne'. In February 1787, he records that 'Mrs. M. to take a table spoonful of Caster oyl twice a week & at night to take the tragacanth draught. Then to omit it for a week or fortnight and begin again.'[8] It would appear that digestive problems plagued the Maskelyne household.

THE NEW MRS MASKELYNE

While Maskelyne's daughter is an almost constant presence in the journals, and her relationship with her father very clear, her mother has a far less dominant place in their pages. The relationship between Sophia and her husband is, therefore, not as easy to read.

What is known of Sophia is that she and her sister Letitia were the daughters of John Pate Rose, who had estates in Northamptonshire and Jamaica. John Rose spent a great deal of time in London in the mid-eighteenth century, where he is thought to have met Martha Henn. Although they never married, they had two children, Letitia (b. 1751) and Sophia (b. 1752), whom John Rose acknowledged as his heirs. He died at the early age of thirty-five and, immediately after his death, Martha changed her name to Mrs Martha Henn Rose. The girls were educated at a boarding school; there is a record from 1762 of the gift of coffee and chocolate to their tutor. In 1773 Sophia came of age and the two sisters were given control of the inheritance which had been put into trust for them. Both lived together, with their mother, in their London property until Martha's death in 1783. In her will she left Cotterstock Hall and its estate to Letitia, while Sophia received a half share of the other properties, which included a house in Northill, Bedfordshire, and an estate at Ramsey, Huntingdonshire. The Jamaican plantation, however, had already been sold.[9] The sisters appear to have divided their time between Cotterstock and a house in London, located in Milman Street, Holborn. In July 1784, nine months after their mother's death, Letitia married the widowed Reverend Sir George Booth (1724–97), a cousin of Nevil Maskelyne, who had inherited the baronetcy of Dunham Massey in 1770 and whose first wife, Hannah, had died in March 1784.[10] Both Maskelyne and Sophia were witnesses at the wedding.[11] One month later, Sophia and Nevil Maskelyne were married at St Andrew's, Holborn. At the time of the wedding she was thirty-two years old, and he was fifty-two. Repaying the earlier favour, George and Letitia Booth were witnesses.[12]

One reason that may account for the sisters' marrying at a mature age, and a relatively short time after their mother's death, was their illegitimacy and how this might have affected their prospects in the London marriage market of the late eighteenth century. Natural daughters of the upper-middle class were able to make socially advantageous marriages, provided that they could bring an appreciable financial benefit to the partnership. Should a settlement not be as generous as expected, marriage arrangements quickly fell through, prompting one wag to comment that, in order for one's natural daughter to make a suitable marriage, the father had better 'scrape all he can together for her'.[13] Given that the Rose girls did not marry until after their mother died, it would appear that, with the Jamaica estate already sold, there was perhaps not enough to settle on them earlier and also allow Martha to continue living in comfort.

A purple brocade sacque in the collection of the National Maritime Museum was very probably worn by Sophia Rose at her wedding to Nevil Maskelyne (Fig. 2). Silks in this period were

NATIONAL MARITIME MUSEUM, ZBA4678

Fig. 2: A purple silk brocade dress, made about 1785, belonging to Sophia Maskelyne and probably first worn at her wedding to Nevil Maskelyne.

costly and the fabric was often purchased at great expense; and, as full-skirted dresses were made of large lengths of uncut silks, it was economically practical to simply unpick a dress from the 1770s and remake it in the fashions of the 1780s. Extant examples, specifically in the Snowshill Collection of the National Trust, certainly support this.[14] We know that Sophia did refurbish this dress, as it was preserved with additional pieces from the bodice, which had been kept to replace any parts that became noticeably worn. From the small collection of clothing that survives from Sophia's wardrobe, we can see that she kept much older pieces, possibly those belonging to her mother, and simply altered them for her own use. A bed gown of double-weave wool and a pale blue quilted-silk petticoat from the first half of the eighteenth century were preserved in the Maskelyne collection. Both had been taken in at the neck and waist to fit a smaller figure than the individual for whom they were originally made. The bed gown, a short jacket worn loosely crossed and tied at the waist, was a popular garment in eighteenth-century England. Although the name implies that it would have been strictly limited to indoor wear, that was not the case, and it was a staple of the female wardrobe, generally worn with a petticoat. It is likely that these two garments were worn together: although the wool is fine, it would have been a foil for the luxury object that is the blue silk quilted petticoat. While still in use in the 1780s, the combination would not by then have been fashionable and was most probably only worn amid family in the home.

The sacque dress was an old-fashioned choice for 1784 and this conservatism may also be reflected in other choices made by the Rose women, including their London address in Milman Street, Holborn. This street was built in the 1680s by William Milman, the son of a coffee-house keeper who made a fortune in the early stock market. (It still exists but was redeveloped in the 1790s and again in the twentieth century, so retains none of its original appearance.) As a neighbourhood, Holborn was skewered by Fanny Burney in her 1778 novel, *Evelina, or a Young Lady's Entrance into the World*. After a financial reverse forces the title character to move there from the fashionable and respectable address of Queen Anne's Gate, the handsome and wealthy Lord Orville asks to pay a social call. She replies: 'O how I changed colour at this unexpected request! – yet, what was the mortification I suffered in answering, "My Lord, I am – in Holborn!"'[15]

Although Holborn was less fashionable than Queen Anne's Gate, it was perhaps not as wretched as Burney has her character believe. Milman Street itself bordered on Lamb's Conduit Fields and the Foundling Hospital, established in 1739, which enjoyed fashionable patronage, including by well-known artists of the day. Handel was one supporter, while Hogarth and other contemporary painters endowed it with a fine collection of their works, which helped make it a venue for affluent, charitable visitation. Throughout the eighteenth century a popular fundraising activity among the London parishes was also the Sunday charity sermon, many of which were regularly advertised in the London papers. The Reverend Sir George Booth was a fixture of these, preaching at St Mary's, Rotherhithe, for the relief of Sailors; Fulham for the Humane Society and St Dunstan's, Stepney, for the relief of poor boys.[16] The Foundling Hospital was one of the most popular venues for Sunday charity sermons and it is probable that the Rose sisters became acquainted with George Booth and his cousin Nevil Maskelyne through attendance at one or more of them: one can at least suggest so, given that Foundling boys were raised for the sea and Maskelyne's work directed towards improvement of navigation, so it is an institution with which he would, at least, have been familiar.

FAMILY LIFE

There is no mention of Sophia's arrival at the Royal Observatory, yet it must have launched a period of considerable adjustment after Maskelyne's preceding bachelor residence there of nearly twenty years. His journal records even more of a change to the new household in June 1785:

> June 25th 1785 at 20 minutes to one in the morning Mrs. Maskelyne was brought to bed of a girl, being her first child. July 26 she was christened Margaret the name of her godmother my sister Lady Dowager Clive. Thos. Kelsall Esq. cousin was the godfather and Mrs. Maskelyne the other godmother.[17]

The baby's name was chosen from Nevil Maskelyne's side of the family, and perhaps it was politically prudent that she was named after her godmother, Lady Clive.

The first recorded image of Sophia was painted in 1786 as a pair to that of her husband (Fig. 3) by Louis van der Puyl, though they are no longer together.[18] She is shown as a matron wearing a green silk dress with sabot sleeves, again a fashionable but conservative choice. On her lap is the infant Margaret and Sophia's left hand with her wedding ring is prominently displayed. It is possible that this is highlighted because this portrait, a pair to the portrait of Nevil Maskelyne in the Royal Society, is also a marriage portrait. However, it may also reflect the fact that her own parents never married and that she is reinforcing her own legitimacy as an adult and that of her daughter. An entry from Maskelyne's account book indicates that it was paid for on 24 May 1786 and that the portrait and frame cost £25 10s.[19] According to Margaret Clive it was 'a very fine likeness of Sophia'.[20]

Compared to her daughter's presence in them, it is extremely difficult to get a sense of Sophia from her husband's journals, although we do glimpse her. She is mentioned both in terms of her health and recipes of hers that are recorded by Nevil, either as of interest or because of use to him, such as 'Mrs. M.'s Burn Ointment'. However, her own voice is largely lost. Certainly, as a genteel woman of the latter part of the eighteenth century, she would have been expected to manage the household and keep her own housekeeping journals.[21] Maskelyne's account books for the 1780s cover both his own accounts (he does mention lending £10 to Mrs Maskelyne), the household expenditures and the Cotterstock revenues and accounts of his cousin George Booth and sister-in-law Letitia.[22]

Fig. 3: A portrait of Sophia Maskelyne, with her young daughter Margaret Maskelyne, by Louis van der Puyl, 1786. It was painted later, but formed a pair with van der Puyl's portrait of Nevil Maskelyne (see p.7).

It would make sense that Sophia would have kept her own household accounts: we assume that she was literate, as she had been educated to a relatively high degree. She was, despite her illegitimacy, a gentleman's daughter, and had been brought up to follow what would have been her expected social role. The extremely brief glimpses of her in the Maskelyne memorandum books may indicate that she had notebooks of her own for housekeeping and possibly for home remedies, but the contents of the larder are recorded by Nevil. This may be part of his almost compulsive need to make observations, which may duplicate Sophia's own work, as a well-stocked larder and well-kept table were a confirmation of successful housewifery. We also know that Sophia and Margaret tended the flower and vegetable garden at Greenwich with the aid of a part-time gardener: both his pay and various seed orders have been recorded.

It is curious that Sophia is such a shadowy domestic presence. Perhaps ensuing generations simply did not keep her housekeeping journals because they were part of the domestic sphere, but also possibly because there was a great deal of repetition between them and her husband's. Perhaps his were retained because of his reputation as the 'great man,' while their descendants did not place as much value on what hers may have had to tell us. She certainly appears to have been a notable presence at the Royal Observatory and is remembered in letters, but details of her domestic role are lacking.

SOPHIA MASKELYNE'S IMAGE – A FASHIONABLE MATRON?

However, since we have lost Sophia's voice on paper, we must turn to the record of her image. In addition to her portrait of 1786 by van der Puyl, recently married and with her infant daughter, there are two further known surviving ones. The first is a portrait miniature by Mary Byrne (Mrs James Green), from July 1803 (Fig. 4).

Again, it forms a pair with her husband's, which was painted by Byrne in November 1801 (Fig. 5): as his memoranda record, they cost £10 each including the gold frames. In his, as in the 1785 canvas by van der Puyl, Nevil again wears clerical dress and so appears very similar. Sophia's appearance, by contrast, has greatly changed: she is shown as a fashionable matron in a red dress, fine muslin fichu and cap, and with a great quantity of pearls used to secure her husband's miniature portrait to her

Figs 4 and 5: Miniature portraits of Nevil and Sophia Maskelyne by Mary Byrne, painted in 1801 and 1803 respectively. The miniature of Sophia includes that of Nevil, worn as jewellery.

dress. Her costume is one that clearly follows the fashion plates of the early nineteenth century, particularly in the cap with a bow on the back, examples of which can be seen in the London fashion plates for summer 1803, as can the strings of pearls.[23]

It is not unusual for a wife to wear a portrait miniature of a member of her family in this period. The use of such miniatures changed from their inception in the sixteenth century, when they were worn under the clothing or concealed in a locket. By the seventeenth and early eighteenth centuries they were used as means of affirming an alliance – whether this was through marriage or attachment or, as commonly seen, a more overtly political connection, such as wearing a miniature of the monarch as a means of declaring loyalty. By the late eighteenth century, the portrait miniature was worn openly as jewellery, which 'maximised the image's display or "advertising" potential, and associated the miniature image with [the possessor's] own body ... In this way, portrait miniatures were part of the theatrical apparatus of the eighteenth-century presentation of self'.[24] A portrait miniature worn openly, as Sophia displays that of Nevil, would also indicate a very affectionate relationship, as can be seen with Elizabeth Foster, confidante of Georgina, Duchess of Devonshire, who wore such a miniature of Georgina around her neck and called it her 'beloved medallion': it is prominently displayed in Angelica Kauffman's portrait of Foster, painted in 1784. Further, by being shown wearing the portrait miniature of Nevil in her own, Sophia Maskelyne is firmly identifying herself as his wife, in much the same way that her prominently displayed wedding ring in the van der Puyl portrait confirms her marital status and her daughter's legitimacy. This overtly gendered, conventional use of Nevil's miniature may also underscore the notion that Sophia had taken refuge in the domestic sphere of a house that was both a workplace and one of reception, since the Astronomer Royal regularly entertained socially high-ranking visitors from Britain and abroad. The one individual who crosses between these two spheres appears to be Margaret, and it may be the case that she was able to do so because she was Maskelyne's sole heir. A late eighteenth-century portrait of Margaret (then in her early teens), by William Owen, shows her in a muslin dress with her hair dressed in a unisex style known as 'spaniel's ears', with the Royal Observatory in the background again included as an attribute of her family status (Fig. 6).

Fig. 6: An oil portrait of Nevil and Sophia Maskelyne's daughter, Margaret, by William Owen, c.1795–98. Margaret is placed within Greenwich Park, with the Royal Observatory appearing in the background.

The final portrait of Sophia (Fig. 7) was done by John Russell in 1804 and also makes a pair with one of Nevil, this time in pastel (Fig. 5, p269).

Fig. 7: A portrait of Sophia Maskelyne, by the pastellist John Russell, 1804. It forms a pair with the 1804 portrait of Nevil Maskelyne illustrated in the previous Case study.

Russell was not only a Royal Academician but a noted amateur astronomer. He did them in gratitude to Maskelyne for his help, both in producing pastel drawings of the Moon and in his Selenographia, 'A Globe representing the Visible Surface of the Moon', which was published in 1797. In his biography of Maskelyne, Derek Howse cites Lady Clive's acerbic pronouncement to Margaret that 'Nobody is pleased with your father's little portrait' – although she seems to have withheld comment on that of Sophia.[25]

In the latter we see a woman who, as in the Byrne miniature, closely followed the fashion plates. She is dressed in a white muslin gown with a blue shawl of light material over one shoulder. She wears a type of turban and has yellow gloves similar in style to those featured in a fashion-plate advertisement of 1802 by Wirgman, a dressmaker in Hanover Street. Either in a further nod to fashion or to preserve the illusion of youth – since she is fifty-two in this portrait – she also wears a fringe of dark false curls, as the natural grey of her hair can clearly be seen along the edge of her turban. Muslin was both affordable and the height of fashion in the late eighteenth and early nineteenth centuries. Although it was easily stained – and the more expensive, lighter muslins were extremely delicate – it was, above all, washable and it took dye very well, owing to the nature of cotton fibres. Jane Austen and her sister Cassandra dyed their old muslins in pastel colours when their whiteness could no longer be maintained.[26] In addition to its wearability, the lightness of muslin meant that it lent itself particularly well to the approximation of classical drapery, as represented in the much-imitated statuary of ancient Greece and Rome.

Although Sophia was clearly a follower of fashion, the Russell portrait perhaps indicates that she was not always discerning enough to appreciate the connotations of certain styles of dress. Although all ages could, and did, wear muslin, white muslin in particular was the preserve of the young unmarried woman, the young matron and the child.[27] However, on older women white muslin had an almost 'baby-fying' effect.[28] Dress historians have drawn parallels between the female characters in popular novels of the late eighteenth and early nineteenth centuries, specifically the work of Jane Austen, and the way in which their dress denotes their character. The figure of Mrs Allen, Catherine Morland's rather immature middle-aged chaperone, who is habitually swathed in white muslin, is characterized by Austen as follows:

> Mrs. Allen was one of that numerous class of females, whose society can raise no other emotion than surprise at there being any men in the world who could like them well enough to marry them. She had neither beauty, genius, accomplishment, nor manner. The air of a gentlewoman, a great deal of quiet, inactive good temper, and a trifling turn of mind were all that could account for her being the choice of a sensible, intelligent man like Mr. Allen. In one respect she was admirably fitted to introduce a young lady into public, being as fond of going everywhere and seeing everything herself as any young lady could be. Dress was her passion. She had a most harmless delight in being fine.[29]

It is not unusual for the time that the portraits of Sophia Maskelyne, spanning the period from shortly after her marriage until seven years before her husband's death, show a woman whose images were only created in conjunction with his (though not vice versa, since there are a few others of him). While his halves of these pairs consistently illustrate his attributes as a man of learning and academic accomplishment – the Royal Observatory is nearly always present, underlining his position as Astronomer Royal – her identity is always defined through her relationship to him. In her early portrait she is the mother of Margaret Maskelyne; in the miniature she is clearly 'labelled', by the inclusion of Nevil's, as his wife. Only in the last is she seen on her own with no further attributes of her status other than as a fashionable matron who, one might infer, 'had a most harmless delight in being fine'.

MARGARET AND SOPHIA

While Sophia may have run the household, it is clear that Margaret helped her father and she appears to have taken on duties as his secretary. As we saw in Chapter 6, it was she who wrote to Henry Andrews in February 1811 to explain that work on the *Nautical Almanac* must stop due to her father's illness. It also appears that instead of Sophia, in her role as widow, it was Margaret who undertook the task of informing family, friends and professional associates of his death. Sophia's widowed sister, Letitia, who still lived at Cotterstock Hall, wrote back to Margaret in mid-February 1811:

> I this day received your letter, and sincerely Condole with your dear Mamma, and yourself, in the melancholy event that has taken place. The only consolation we can all have is the reflecting on the virtuous life of our late dearly beloved friend, and in considering that he was much advanced in years. Indeed my dear, his departure has been just as I pictured to myself it would be, from his charming serene calmness, and dutiful placid & composed Temper; and in such cases we can have no doubt but the dear Creature is happy. I must say that unless it was by your dear Papa's desire, I could have wished his dear remains might have been brought to Cotterstock: as we might then all have been likely to have kept together with your dear Uncle [George Booth, who had died in 1797] & our dear Parents. But in this Matter your dear Mamma must be the best judge as I am certain she will ever study what would have given my late dear Brother the greatest pleasure – I shall long my dear to see you both and as soon as I get my Mourning made, & some for my Domesticks, I shall hope to be in London which I trust will be in the course of a fortnight. Would it not be proper as we are Patrons of Cotterstock Church: for it to be put into Mourning on this distressing occasion?

A postscript to the letter reveals that it was Margaret who was clearly responsible for informing the family and handling all correspondence: 'I feel much for your dear Aunt Clive when you write to her pray remember to send my affectionate Condolence to her on our heavy loss. I shall be anxious to hear from you again'.[30]

Margaret also dealt with Joseph Banks who, eager to install John Pond as the next Astronomer Royal into his accommodation at the Royal Observatory, wrote to Sophia and Margaret to remind them that they must soon leave. A copy of the reply which most certainly came from Margaret no longer exists, though Banks' subsequent apologetic letter addressed to her does. It would appear that, even before her father's death, Margaret was viewed as his intellectual heir: she certainly played an active role in the dissemination of her father's legacy, maintaining a correspondence with Samuel Vince regarding the publication of Nevil's papers and organizing the information for Banks to forward to the French Academy for Maskelyne's obituary. The extant correspondence from his lifetime also indicates a close relationship between father and daughter. He oversaw her education and she developed a talent in watercolour and drafting that he no doubt encouraged. As mentioned in Case study F, one of her attempts at creating a panorama of the view of Greenwich and London from the Royal Observatory suggests the use of a camera lucida to create it.

In May 1811, Nevil Maskelyne's library and Margaret's camera lucida were sold. In a scathing and potentially libellous article by Olinthus Gregory against Joseph Banks, it was stated that Sophia had offered her husband's library for sale to the government for the use of the new Astronomer Royal. Gregory then wrote of 'how Banks persuaded the Visitors to decline. When the library was sold at auction ... agents employed by Sir Joseph snapped up those books which he thought the most valuable', implying that the widow was thereby swindled out of her just dues.[31]

The only inference one can draw from this are either that Sophia hoped for a higher price by selling the library as a whole to the government, or that Banks spoke against it because he had eyes on such choice personal pickings. However, it is unlikely that Gregory, who would have known Sophia (and that she was not without significant means) was aiming to damage anyone but Banks. Although Howse acknowledges that 'no confirmation of these scurrilous allegations has been found', it is assumed that Gregory felt there was enough truth in them – of Banks as swindler and Sophia as victim – to be plausible within popular currency at the time.

Maskelyne bequeathed Purton Down to his widow and the rest of his estate to Margaret, including Basset Down in Wiltshire. Both women took up residence at Basset Down and in 1819, at the age of thirty-three, Margaret married Anthony Mervyn Story. The courtship was not a smooth one and here we see a rather different face of Sophia Maskelyne. Story paid a visit to the Maskelynes, with 'the full intention of being married in the course of 5 or 6 weeks'. This he related to his mother, but it appears from her correspondence that Sophia steered him on quite a different course:

> Now from what Mrs. M unfortunately said to him & from the doubts which she starts, that never before struck his mind, resulting her doubting, whether their united income would enable him to keep up both Establishments in town and Basset Down and to provide everything for her daughter in the same stile of propriety in which he knew her to have ever been accustomed, he seemed to have been frightened & scared at the suggestions she had put into his head, & at the trouble of getting a House in town, & furniture, carriage, & the numberless new duties that would /he fancied/ devolve upon him to fulfil properly, & talks now of being a Bachelor for life.[32]

The letter goes on to quantify the cost of living to maintain an establishment both in London and Wiltshire, and expresses the opinion that Story does not have an income to do this. Further, Story's brother expresses disbelief when, upon enquiry, both he and Story discover that Sophia is correct in estimating the costs of living and that her sums are significantly higher than those Story had calculated. Her knowledge of the cost of maintaining a house, the various other related necessities and, indeed, the basic costs of housekeeping, point towards a thorough knowledge on her part of domestic accountancy.

There are also further hints of the role that Sophia played in the Maskelyne household. In 1816 she is mentioned in a lawsuit for breach of contract over the promise of marriage between a Mr Lancy and Miss Hunter. According to the account of this, 'The defendant was a widower, and a Gentleman of considerable fortune, age 51 to whose the daughters the Plaintiff had been governess during their mother's life time, to which situation she was recommended by the lady of Dr Maskelyne, Astronomer Royal'.[33] This demonstrates a typical exploitation of social and familial networks, in which Sophia both drew on her own connections and assisted her husband's; for the plaintiff, Miss Hunter, was the daughter of a master of mathematics at the Royal Hospital School, Greenwich.

Sophia died in February 1821. Her death appeared simply in the papers: 'On Tuesday last at her seat, Basset Down House, in her 69th year, Sophia, relict of the Rev. Nevil Maskelyne DD, late Astronomer Royal.'[34] She died intestate, which might at first seem unusual but perhaps ties in with her passing of responsibilities to Margaret. Moreover, Sophia only had one child, who had been adequately provided for through Maskelyne's will and who was over the age of minority, so it was perhaps 'less essential to provide for family members and to preserve domestic stability'.[35] The letters of Administration were granted to Margaret Story by consistory court later in 1821, leaving her the sole heir of both her parents. Sophia's sister Letitia died in September 1823 at Cotterstock, but it is possible that, as key responsibilities after the death of Nevil Maskelyne devolved to Margaret, it was her role to ensure that Letitia was given something in remembrance of her sister.

The sense that we get of the Royal Observatory and the Maskelyne family, particularly Sophia, is that women's roles in the early nineteenth century did move into a more restrictive domestic sphere than had previously been the case, although perhaps not as completely separated by gender as earlier historians have been inclined to believe. One factor in creating this perceived division between the domestic and public spheres, and their stringent gendering, was the increasing importance of the popular and widespread Evangelical movement. This stressed the role of women in the home, which in turn helped encourage an increasingly conservative society in the nineteenth century. Emphasis on the domestic virtues is lauded in popular novels of the period – such as *Marriage* (1818) by Susan Ferrier – which advocated that the role of women was to make the home both a place of welcome and of peace. Jane Austen's slightly earlier novel, *Mansfield Park* (1814), contrasts a very domestic and quiet Fanny Price in the role of heroine, with a more decadent, socially active Mary Crawford.

It appears that while the household was together in the Royal Observatory, the family played to these expected roles. Certainly Nevil Maskelyne had the outward-facing position and his voice has dominated those of the others, not least in the degree to which they can still be heard today. Margaret's is there to some extent as his only child and heir but Sophia, her mother, is now barely audible, owing both to circumstance and what would seem her own view of the part she played.

FURTHER READING

In *The Gentleman's Daughter* (Yale University Press, 2003) Amanda Vickery provides a comprehensive overview of women's roles in both public and private spheres in Britain during the eighteenth and nineteenth centuries. Vickery has used diaries and correspondence to map the way in which women developed and exploited social networks – this is a particularly useful resource in trying to make sense of the Maskelyne family's domestic life in the Royal Observatory.

For a history of dress and fashion during this period, Jane Ashelford, *The Art of Dress: Clothes and Society, 1500–1914* (Harry N. Abrams, 1996) gives a good overview of changing taste, shopping habits, and fashion publications in eighteenth-century London. Drawing almost exclusively on the collections of the National Trust, Ashelford very usefully compares the idealized image of how costume should appear with the actual garments themselves.

In *Psychosocial Spaces: Verbal and Visual Readings of British Culture, 1750–1820* (Wayne State University, 2003), Steven J. Gores discusses the gendered meanings of dress and accessories during the eighteenth century, particularly in respect to the miniature portrait.

CODA: A LIFE WELL LIVED

In Derek Howse's biography, Maskelyne is styled as 'the seaman's astronomer'. This, of course, underlines the essential connection between his astronomy and maritime navigation. He instigated the *Nautical Almanac*, still the essential tool of nautical astronomy, he worked to make both the lunar-distance and timekeeping methods of finding longitude at sea comprehensible and affordable, and he played an important role in the organization of voyages of exploration. The practical bent of his work was his motivation, and the belief in 'the usefulness of his life of learning' was literally set in stone on his memorial tablet in the Church of St Mary at Purton, Wiltshire.[1]

However, in his lifetime, only a small proportion of seamen would have been able – or, indeed, would have felt the need – to make use of the *Nautical Almanac*. Few, even in 1811, sailed on a ship supplied with a chronometer. The long-term legacy of the eighteenth-century effort to solve 'the longitude problem' had yet to play out, although the techniques were beginning to be applied to navigation and, perhaps more importantly, to charting. It was because those charts were based on the data available in the *Almanac*, which relied on Maskelyne's observations, that the meridian of the Royal Observatory at Greenwich became increasingly important. In the late eighteenth century, the Greenwich Meridian became the Prime Meridian for Ordnance Survey maps and Admiralty charts. As the latter grew in coverage and became widely used, so Greenwich eventually became the international reference for longitude and time.

Yet, despite this legacy, Maskelyne might better be described as 'the astronomer's astronomer'. As we have seen, he was highly regarded by other astronomers in Britain, Europe and beyond, and it is in histories of astronomy rather than navigation that he has found his place. Those with whom he worked most closely in relation to maritime voyages were astronomers sponsored by the Board of Longitude – his former assistants and computers – who carried out their observations at sea and on land to help pursue the central projects of eighteenth-century astronomy. Maskelyne's most significant legacy, from this point of view, was his transformation of the Royal Observatory and his published *Greenwich Observations*, which were essential to later improvements of lunar and planetary theory.

The seventh Astronomer Royal, George Biddell Airy (1801–92), who is often given credit for creating an efficient, factory-like system at Greenwich, owed much to his predecessor-but-one.[2] Although he considered Maskelyne's 'systems of observation' much inferior to those of the sixth Astronomer Royal, John Pond, who had the benefit of several assistants, it was Maskelyne who refocused efforts on navigation and ensured that the Observatory and its work were always carefully managed.[3] Patrick Kelly's description of the 'steady perseverance' and 'efficiency' of Maskelyne's approach to his work, directed 'to the practical purposes of life', was undoubtedly one that might have applied to Airy too.[4]

The 'life well lived' that heads this brief section also appears on Maskelyne's memorial tablet.[5] It might be taken to refer not just to his astronomical work but also his sense of public duty, played out through his role as a Commissioner of Longitude, council member of the Royal Society and, on occasion, advisor to government on matters technical and mathematical. Maskelyne's sense of duty extended, as we have seen, to generosity in his correspondence with those seeking information and advice, and assistance for those who sought or were under his patronage. It is clear, though, that he did not suffer fools or timewasters gladly and that, like other Commissioners of Longitude, he was impatient with those who did not deal with the Board straightforwardly. Nevertheless, to quote Howse, the view that Maskelyne was spiteful, jealous or patronizing to Harrison

> was certainly not one that was held generally in his own day, nor is it in any way justified by today's research.... There is no evidence whatsoever that Maskelyne at any time abused his position as a public servant in order to further his own ends, still less to line his own pocket.[6]

The inscription on the memorial tablet was probably composed by Maskelyne's daughter, Margaret, who wrote to Samuel Vince for advice in its composition. He suggested that it note how Maskelyne 'discharged most faithfully all [his] Duties with the highest credit to himself and to his Country', a sentiment with which Margaret undoubtedly agreed, and the final legend bears an echo of the following from Vince's letter:

> His mildness of Manner and gentleness of Disposition, gained him the affection of his Friends, and the Esteem of all who had the Happiness of his Acquaintance. He died lamented as he had lived beloved; and by his Death the public lost one of its most distinguished and faithful servants.

However, she downplayed Vince's rather florid claims regarding Maskelyne's religious expression, which perhaps represented his own stance more than her father's:

> Conversant with the great works of the creation, he was led to the acknowledgement of, and had the highest veneration for, their Author; and as a Minister of the Gospel, his Belief in revealed Religion was firm, and his Practice conformable to his Profession. The sincerity of his Trust in God, was manifest through the whole Course of his Life; but more strongly seen in his calm and patient Resignation to the will of his Maker, when he found his End approaching.[7]

In the final version, Maskelyne's principal act of worship was his work:

> In the cemetery of this church are deposited the remains of

NEVIL MASKELYNE D.D. F.R.S. Astronomer Royal, and whether you look at the simplicity of his way of life or the kindness of his heart or the usefulness of his life of learning, this man is worthy of being publicly mourned since he worshipped the Great Creator by formulating laws of Nature: Virtuous without pretence, he demonstrated goodness in his work, ultimately trusting not in himself, but in Christ, to the bosom of the Eternal Father he rendered a life well lived in the certain hope of the reward to come.[8]

ENDNOTES

ABBREVIATIONS IN REFERENCES

CUL:	Cambridge University Library
MP:	Nevil Maskelyne Papers (REG09/000037), National Maritime Museum, Caird Library
NMM:	National Maritime Museum
RAS MSS Pigott:	Royal Astronomical Society Manuscripts, Nathaniel Pigott papers
RGO:	Royal Greenwich Observatory Archives, Cambridge University Library (including Board of Longitude papers, RGO/14, and Maskelyne papers, RGO/4)

INTRODUCTION

1 These sums are equivalent to more than £1500 today, a sum that in the 1780s was, the National Archives' currency converter tells us, worth at least two horses or 159 days' wages for a labourer. The costs are recorded in one of Maskelyne's account books and given in Derek Howse, *Nevil Maskelyne: The Seaman's Astronomer* (Cambridge: Cambridge University Press, 1989), pp. 144–45.

2 This account of Maskelyne's early life is mostly based on Howse, *The Seaman's Astronomer.*

3 The text is transcribed in Howse, *The Seaman's Astronomer*, pp. 214–24. The original is at CUL, RGO 4/320 and online at Cambridge Digital Library, beginning at http://cudl.lib.cam.ac.uk/view/MS-RGO-00004-00320/35.

4 Nevil Maskelyne, 'Books bought or given me since my first coming to the University Novr. 5 1749', Wiltshire Record Office, Story Maskelyne of Purton, MS 1390/13.

5 Sten G. Londberg (ed.), *Bengt Ferrner: Resa in Europa, en astronom, industriespion och teaterhabitue genom Denmark, Tyskland, Holland, England och Italien – 1758–1762* (Uppsala, 1956), entry for 29 April 1760 (translation supplied by Jacob Orrje).

6 Many of the objects have been catalogued online by the NMM: http://collections.

rmg.co.uk/, see e.g. items ZBA4664–84. The manuscripts (NMM MP) have been digitized at Cambridge Digital Library as part of the Board of Longitude collection http://cudl.lib.cam.ac.uk/collections/longitude (starting at http://cudl.lib.cam.ac.uk/view/MS-REG-00009-00037/1).

7 'The Board of Longitude, 1714–1828: science, innovation and empire in the Georgian world', funded by the Arts and Humanities Research Council: see http://blogs.rmg.co.uk/longitude/.

CHAPTER 1

1 Dava Sobel, *Longitude: The True Story of a Lone Genius Who Solved the Great Scientific Problem of His Time* (New York: Walker, 1995; London: Fourth Estate, 1996); *Longitude* (Granada Productions and the A&E Network for Channel 4, first broadcast in the UK, January 2000).

2 Sobel, *Longitude*, p. 111.

3 There are three manuscript copies of the Harrison 'Journal', one of which is at the State Library of New South Wales, now online at http://cudl.lib.cam.ac.uk/new/MS-H-17809/1.

4 The text is transcribed in Derek Howse, *Nevil Maskelyne: The Seaman's Astronomer* (Cambridge: Cambridge University Press, 1989), pp. 214–24, while the original manuscript is at CUL RGO4/320 and online at Cambridge Digital Library http://cudl.lib.cam.ac.uk/view/MS-RGO-00004-00320/35.

5 See http://cudl.lib.cam.ac.uk/view/MS-RGO-00004-00320/36; there seems to be no clear reason for this discrepancy, although the words 'stood the 3^{d}' appear to be written over the deleted 'obtained [blank]th', which may indicate that Maskelyne was unsure of the number.

6 May it continue in perpetuity.

7 'Some account of the Rev. Nevil Maskelyne, D.D., F.R.S., and Astronomer Royal', *European Magazine, and London Review* (June 1805), 407–08.

8 [Patrick Kelly], 'Maskelyne, Nevil', in *The Cyclopaedia; or, Universal Dictionary of Arts, Sciences, and Literature,* ed. by Abraham Rees, vol. 22 (London, 1812), pp. 416–22.

9 [Thomas Young], 'Maskelyne (Nevil)', *Supplement to the Encyclopaedia Britannica*, vol. 5 (Edinburgh, 1824), pp. 313–15.

10 Jean Baptiste Joseph Delambre, 'Biographical Account of the Rev. Nevil Maskelyne', *Annals of Philosophy,* 50 (1813), 401–14.

11 Lalande to Maskelyne, 10 August 1787, NMM MP http://cudl.lib.cam.ac.uk/view/MS-REG-00009-00037/722. The full text of the postscript is 'Je vous recommande toujours M. de Lambre, il fait bien honneur a vos observations dans ce pays, vous etes pour lui le dieu de l'astronomie; et il en est le soutien en France' (I again

commend M. de Lambre to you. He pays much tribute to your ideas in this country. For him you are the god of astronomy. He is its [astronomy's] mainstay in France).

12 Samuel Vince to Margaret Maskelyne, 7 December 1811, NMM MP http://cudl.lib.cam.ac.uk/view/MS-REG-00009-00037/620.

13 Samuel Vince to Margaret Maskelyne, 15 December 1811, NMM MP http://cudl.lib.cam.ac.uk/view/MS-REG-00009-00037/623.

14 Samuel Vince to Margaret Maskelyne, 21 January 1812, NMM MP http://cudl.lib.cam.ac.uk/view/MS-REG-00009-00037/624.

15 Samuel Vince to Margaret Maskelyne, 2 October 1811, NMM MP http://cudl.lib.cam.ac.uk/view/MS-REG-00009-00037/618.

16 Stephen Rigaud to Margaret Maskelyne, 24 February 1832, NMM MP http://cudl.lib.cam.ac.uk/view/MS-REG-00009-00037/632.

17 Jean Sylvain Bailly to Maskelyne, 21 May [1775?], NMM MP http://cudl.lib.cam.ac.uk/view/MS-REG-00009-00037/728.

18 John Bentley, *A Historical View of the Hindu Astronomy* ... (London: Smith, Elder & Co, 1825), p. xxv.

19 Robert Grant, *History of Physical Astronomy From the Earliest Ages to the Middle of the Nineteenth Century* (London, 1851), p. 208, p. 490.

20 [Augustus De Morgan], 'Maskelyne', in *The Gallery of Portraits: With Memoirs*, ed. by A. T. Malkin, vol. 6 (London, 1836), pp. 20– 24.

21 A. M. C. [Agnes Mary Clerke], 'Nevil Maskelyne', *Dictionary of National Biography* (Oxford, 1893).

22 'An Account of John Harrison', *The European Magazine, and London Review* (October 1789), 235–36.

23 'Harrison, John', in *General Biography; or, Lives, Critical and Historical, of the Most Eminent Persons of all Ages, Countries, Conditions, and Professions*, ed. by John Aikin, Thomas Morgan and William Johnston, vol. 5 (London, 1804), pp. 63–65.

24 'Harrison', in *The Gallery of Portraits* ..., vol. 5 (London, 1835), pp. 153–56.

25 Johan Horrins [John Harrison], *Memoirs of a Trait in the Character of George III* (London, 1835), see particularly the extensive footnotes to the preface, from which these quotes are taken.

26 Edward J. Wood, *Curiosities of Clocks and Watches: From the Earliest Times* (Richard Bentley: London, 1866).

27 Jonathan Betts, *Time Restored: The Harrison Timekeepers and R.T. Gould, the Man Who Knew (Almost) Everything* (Oxford: Oxford University Press, 2011), p. 267.

28 Humphrey Quill, *John Harrison: The Man who Found the Longitude* (London: John Baker, 1966), p. 175.

29 Quill, *Harrison*, Author's Preface, p. 173, p. 117.

30 J. A. Bennett, 'Horology, history and Harrison [Review of Andrewes (ed.), *The Quest for Longitude*], *Journal of Antiquarian Horology* 23 (1997), 451–56, esp. p. 454.

31 Howse, *The Seaman's Astronomer*. 'The Nautical Almanac Bicentenary' exhibition was at the NMM in 1967.

32 Bennett, 'Horology, history and Harrison', p. 453.

33 Several letters and Bennett's responses appeared in *Antiquarian Horology*, 23 (1997), pp. 558–61, p. 24 (1998), pp. 77–82, 169, 269, 380.

34 David Philip Miller, 'The "Sobel Effect": The Amazing Tale of How Multitudes of Popular Writers Pinched all the Best Stories in the History of Science and Became Rich and Famous while Historians Languished in Accustomed Poverty and Obscurity, and how this Transformed the World. A Reflection on a Publishing Phenomenon', *Metascience*, 11 (2002), 185–200; John Hedley Brooke, 'Presidential Address. Does the History of Science have a Future?', *British Journal for the History of Science*, 32 (1999), pp. 1–20; Simon Schaffer, 'Our Trusty Friend the Watch', *London Review of Books*, 18 (31 October 1996), 11–12; John Gascogine, '"Getting a Fix": The *Longitude* Phenomenon', *Isis*, 98 (2007), 769–78.

35 Miller, 'The "Sobel Effect"', p. 195.

CHAPTER 2

1 Derek Howse, *Nevil Maskelyne: The Seaman's Astronomer* (Cambridge: Cambridge University Press, 1989), p. 26; Royal Society Archives (hereafter 'RS') AB/1/2/1/24; CMO/4/112; CMO/4/115; CMO/4/118; CMO/4/120; CMO/4/121; CMO/4/133; MM/10.

2 British Library Add MS 4320, f. 83.

3 *The Gentleman's Diary, or the Mathematical Repository* (London, 1758), p. 23; Ibid. (London, 1761), p. 41.

4 Benjamin Martin (ed.), *The General Magazine of Arts and Sciences*, 11 (1759), pp. 1–2.

5 Ibid., pp. 29, 30, 54, 78, 125, 156, 183, 204, 230, 251, 330, 356, 380, 428, 453; see also Ibid., p. 229.

6 Ibid., p. 568.

7 RS CMO/4, p. 347.

8 Royal Astronomical Society (hereafter RAS) MSS Pigott, pp. 84–103.

9 Howse, *The Seaman's Astronomer*, p. 245, note 18.

10 Martin, *General Magazine*, p. 561.

11 RAS MSS Pigott, 84.

12 Captain Charles Haggis, Journal of the *Prince Henry*, British Library IOR/L/MAR/B/325G.

13 The dates in the ships' journals and the records kept by Waddington are not always consistent, in part because of the way dates were reckoned at sea (i.e. from noon to noon).

14 Captain George Kent, journal of the *Hawke*, British Library IOR/L/MAR/B/390B, 2 Feb.

15 Captain Haggis also notes this ship in his journal.

16 Captain Haggis's journal notes sermons by Maskelyne on 15 and 22 February.

17 The letter to Pigott says 9 March: Captain Kent's journal says 11 March. The cheers and good wishes are recorded in the latter.

18 Captain Haggis's journal records that the ship was the *Mercury*, Captain Harold.

19 Maskelyne's lunar observations are recorded in his journal, op. cit., and generally in Captain Haggis's journal.

20 Captain William Webber, journal of the *Oxford*, British Library IOR/L/MAR/B/588D.

21 RAS MSS Pigott, p. 85.

22 *London Evening Post*, 5295 (6–8 October 1761).

23 *Whitehall Evening Post or London Intelligencer*, 2428 (6–8 October 1761); *Public Ledger or The Daily Register of Commerce and Intelligence*, 546 (9 October 1761).

24 *Public Ledger*, 568 (4 November 1761).

25 Waddington's contributions to Benjamin Martin's publications now come from that address, Benjamin Martin, *Miscellaneous Correspondence*, Vol. 4 (London, 1761), pp. 965, 1100.

26 I am grateful to Alexi Baker for checking her transcription of the minutes of the Board, now available online at http://cudl.lib.cam.ac.uk/collections/longitude.

27 RAS MSS Pigott, p. 86.

28 RAS MSS Pigott, p. 87.

29 RAS MSS Pigott, p. 97. Through misunderstanding the date of this letter, the RAS Archivist in about 1975 regrettably misplaced it in the chronological sequence (*mea culpa*).

30 RS JBC/24, pp. 258–9; JBO 25/29.

31 RS JBO/25/ 31, 33, 42, 53.

32 Martin, *Miscellaneous Correspondence*, pp. 879–81, 895–7.

33 Ibid., pp. 911–12.

34 RAS MSS Pigott, p. 88.

35 RS CMO/4, pp. 342–4; unfortunately no copy of Waddington's 'petition' can be found. For Ferguson's application and the subsequent procedure, see John R. Millburn, *Wheelwright of the Heavens: the Life and Work of James Ferguson, FRS* (London: Vade-Mecum Press, 1988), pp. 143–44.

36 RS CMO/4, pp. 345–8.

37 RS JBC/24, pp. 583–9.

38 *Lloyd's Evening Post and British Chronicle*, 881 (4 March 1763).

39 Waddington sends Pigott a copy on 18 April 1763, RAS MSS Pigott, p. 90.

40 Robert Waddington, *A Practical Method for Finding the Longitude and Latitude of a Ship at Sea, by Observations of the Moon* (London, 1763), pp. i–ii.

41 *The Monthly Review or Literary Journal*, 29 (October 1763) 307–10.

42 RAS MSS Pigott, p. 89.

43 Jérôme Lalande, *Journal d'un Voyage en Angleterre*, ed. Hélène Monod-Cassidy (Oxford, 1980), p. 26.

44 RAS MSS Pigott, p. 90.

45 R. Waddington, 'A calculation of the solar eclipse', in Martin, *Miscellaneous Correspondence*, pp. 1098–1100.

46 Robert Waddington, *A Supplement to the Treatise for Finding the Longitude* (London, 1764).

47 RAS MSS Pigott, p. 91.

48 RAS MSS Pigott, p. 92.

49 Robert Heath, *The Palladium-Supplement, Enlarged* (London, 1764), p. 19.

50 Howse, op. cit., p. 42.

51 RAS MSS Pigott, p. 93.

52 RAS MSS Pigott, p. 94.

53 RAS MSS Pigott, p. 95.

54 RAS MSS Pigott, p. 96.

55 RAS MSS Pigott, p. 98.

56 Howse, op. cit, p. 43; H.W. Dickinson, *Educating the Royal Navy: Eighteenth- and Nineteenth-century Education for Officers* (London and New York: Routledge, 2007), p. 36. For the movement of Waddington's belongings during this episode, see TNA ADM 106/1138/82, 106/1138/84, 106/1161/29.

57 RS CMO/5, pp. 172–7, 186, 198.

58 RS CMO/5, p. 264.

59 CUL manuscript RGO 14/5, pp. 205–6, confirmed minutes of the Board of Longitude, 1737–1779, http://cudl.lib.cam.ac.uk/view/MS-RGO-00014-00005/209-10.

60 Robert Waddington, *A Description of Instruments Used by Surveyors* (London, 1773).

61 Robert Waddington, *Of the Reflecting Sextant, and Its Use at Sea* (London, 1775?).

62 Robert Waddington, *An Epitome of Theoretical and Practical Navigation* (London, 1777).

63 Robert Waddington, *The Sea Officers Companion, Being an Appendix to Waddington's Navigation* (London, 1778).

64 *Public Advertiser*, 10625 (17 November 1768).

65 RAS MSS Pigott, p. 99.

66 RAS MSS Pigott, p. 100.

67 RAS MSS Pigott, p. 101.

68 William Mountaine, *The Practical Sea-gunner's Companion* (London, 1781), with Robert Waddington, *A Supplement to the Practical Sea-gunner's Companion, Containing the Elementary Part of Gunnery; Together with a Variety of Practical Examples*; Reuben Burrow, *A Restitution of the Geometrical Treatise of Apollonius ... Also the Theory of Gunnery ...* (London, 1779), p. xxvi.

69 RAS MSS Pigot, p. 103.

70 TNA PROB 11/1058/85.

Case study B

1 Transits of Venus occur in pairs, separated by more than a century. Thus there were transits in 1631 (which no one is known to have observed) and 1639 (observed by Jeremiah Horrocks and William Crabtree), 1761 and 1769, 1874 and 1882, 2004 and 2012.

2 'Autobiographical notes', CUL RGO 4/320 http://cudl.lib.cam.ac.uk/view/MS-RGO-00004-00320/35.

3 Royal Society Council Minutes 4, 10 July 1760.

4 This was how it was glossed by the Royal Society when they successfully asked the Admiralty for a man-of-war to transport the Bencoolen expedition, quoted in Howse, *The Seaman's Astronomer*, p. 25.

5 Nevil Maskelyne, 'A Proposal for Discovering the Annual Parallax of Sirius', *Philosophical Transactions*, 51 (1760), 889–95.

6 Ibid., p. 889.

7 Ibid., pp. 891–92.

8 Nevil Maskelyne, 'An Account of Observations made on the Mountain Schehallien, for Finding its Attraction', *Philosophical Transactions*, 65 (1775), 500–42.

CHAPTER 3

1 James Bradley, 'A letter to the Right Honourable George Earl of Macclesfield concerning an apparent motion observed in some of the fixed stars', *Philosophical Transactions*, 45 (1748), 1–43, pp. 5–6.

2 Quoted in Kurt Møller Pedersen and Peter de Clercq (eds), *An Observer of Observatories: The Journal of Thomas Bugge's Tour of Germany, Holland and England in 1777* (Aarhus: Aarhus University Press, 2010), pp. 128, 160.

3 William Ludlam, *Astronomical Observations Made in St John's College, Cambridge, in the Years 1767 and 1768: With an Account of Several Astronomical Instruments* (Cambridge, 1769), p. 40.

4 James Short, 'An Account of the Transit of Venus over the Sun ...', *Philosophical Transactions*, 52 (1761), 178–82, p.178.

5 Nathaniel Bliss, 'Observations on the Transit of Venus over the Sun ...', *Philosophical Transactions*, 52, 173–7, p. 175.

6 James Short, 'Second paper concerning the parallax of the Sun determined from the observations of the late transit of Venus ...', *Philosophical Transactions*, 53 (1763), pp. 300–45, pp. 327–29.

7 Nevil Maskelyne, 'Observations on a Clock of Mr. John Shelton, made at St Helena', *Philosophical Transactions*, 52 (1762), 434–43, p. 437.

8 James Short, 'An Account of Mr. Mason's Paper, Concerning the Going of Mr. Ellicott's Clock at St Helena', *Philosophical Transactions*, 52 (1762), 540–42, p. 541.

9 Nevil Maskelyne, 'Astronomical Observations made at the Island of St Helena', *Philosophical Transactions*, 54 (1764), 348–86, p. 381.

10 Nevil Maskelyne, *Tables for Computing the Apparent Places of the Fixt Stars and Reducing Observations of the Planets* (London, 1774), p.iii.

11 Nevil Maskelyne, *Astronomical Observations Made at the Royal Observatory at Greenwich from the Year MDCCLXV to the Year MDCCLXXIV* (London, 1776), p.iii.

12 Quoted in Derek Howse, 'The Tompion Clocks at Greenwich and the Dead-Beat Escapement, Part 1 – 1675–1678, *Antiquarian Horology*, 7 (1970) 18–34.

13 Royal Society Council Minutes, 5 (12 April 1765).

14 *Miscellaneous Works and Correspondence of the Rev. James Bradley*, ed. S.P. Rigaud (Oxford, 1832), p. 84.

15 Royal Society Journal Book, 24 (9 June 1763).

16 Royal Society Council Minutes, 5 (8 November 1764).

17 Nevil Maskelyne, *Astronomical Observations Made at the Royal Observatory at Greenwich* (London: 1776), p.i.

18 Nevil Maskelyne, *An Account of the Going of Mr John Harrison's Watch, at the Royal Observatory, from May 6th, 1766, to March 4th, 1767* (London: 1767).

19 M. De la Lande, 'Extract of a letter from M. De la Lande ... to the Rev. Mr. Nevil Maskelyne', *Philosophical Transactions*, 52 (1762), 607–10, p. 609.

20 John Winthrop, 'Extract of a Letter from Mr. John Winthrop, Professor of Mathematics in Cambridge, New England, to James Short', *Philosophical Transactions*, 54 (1764), 277–8.

21 James Short, 'Observation of the Eclipse of the Sun ...', *Philosophical Transactions*, 54 (1764), 107–105 [*sic*], p. 107.

22 James Ferguson, 'Observation of the Eclipse of the Sun ...', *Philosophical Transactions*, 54 (1764), 108–13, p. 109, p. 110.

23 Nathaniel Bliss, 'Observations on the eclipse of the Sun...', *Philosophical Transactions*, 54 (1764), 141–44, p. 142.

24 Royal Society Miscellaneous Manuscripts, 244/3–8.

Case study C

1 Thomas Bugge, *An Observer of Observatories: The Journal of Thomas Bugge's Tour of Germany, Holland and England in 1777*, eds Kurt Møller Pedersen and Peter de Clercq (Aarhus: Aarhus University Press, 2010) p. 163.

2 'Qualities to be required for an Assistant May 19. 1787', Memorandum Book 1782–88, Story Maskelyne of Purton papers, Wiltshire and Swindon Archives, WRO1390/2d.

3 Quoted in John Evans, *Juvenile Tourist* (James Cundee, London, 1810), pp. 333–5.

4 David Kinnebrook to his father, 20 August 1794, CUL RGO 35/99 (photocopies: whereabouts of the originals unknown).

5 Kinnebrook to his father, 23 June 1794, CUL RGO 35/97.

6 Kinnebrook to his father, 20 August 1794.

7 Mary Herschel, *Memoir and Correspondence of Caroline Lucretia Herschel* (New York: Appleton & Co., 1876), p. 41.

8 Ibid., p. 89.

9 Kinnebrook to his father, 20 August 1794.

10 Kinnebrook to his father, 15 October 1794, CUL RGO 35/101.

11 Quoted in Howse, *The Seaman's Astronomer*, p. 235. See Samuel Vince to Margaret Maskelyne, 8 July 1812, NMM MP http://cudl.lib.cam.ac.uk/view/MS-REG-00009-00037/628 for his suggestions for the inscription.

CHAPTER 4

1 In addition, it would be necessary to spend 2s 6d on a one-off purchase of a book of reference tables called *Tables Requisite* and to have a sextant or octant, which the ship would probably have carried anyway.

2 Minutes of the Board of Longitude, 30 May 1765, CUL RGO 14/4, f. 93 http://cudl.lib.cam.ac.uk/view/MS-RGO-00014-00005/95.

3 For example, see a manuscript written by Maskelyne entitled 'Rules for computing the Immersion of a Star behind or the Emersion of a Star from the Moon's Limb', CUL RGO 4/216.

4 These are listed in Maskelyne's 'Diary of Nautical Almanac work', CUL RGO 4/324 f. 21v http://cudl.lib.cam.ac.uk/view/MS-RGO-00004-00324/440.

5 Ibid http://cudl.lib.cam.ac.uk/view/MS-RGO-00004-00324/1.

6 Copies of the letters are in CUL RGO 35/97-128.

7 See Chapter 10 in Thomas Middleton, *The History of Tideswell Grammar School* (Hyde, 1933).

8 A copy of which is held at the NMM's Caird Library, MRF/184 (the original is at the Wiltshire Record Office 1390/2).

9 'On the death of the Rev. Mr. J. Edwards', *Shrewsbury Chronicle*, 13 (27 March 1784), p. 3.

10 There are various letters from Mary Edwards among the 'Petitions and memorials' sent to the Board of Longitude (CUL RGO 14/11). The one explaining that she had been computing while her husband was still alive is dated 5 Dec. 1811, RGO 14/11, f. 143 http://cudl.lib.cam.ac.uk/view/MS-RGO-00014-00011/255.

11 For example, see the petition recorded in the Board of Longitude minutes for 7 Dec. 1793, RGO 14/6, pp. 207–08 http://cudl.lib.cam.ac.uk/view/MS-RGO-00014-00006/353.

12 Minutes of the Board of Longitude 5 March 1812, RGO 14/7 f. 158 http://cudl.lib.cam.ac.uk/view/MS-RGO-00014-00007/320.

13 T.C. Hansard, *The Parliamentary Debates*, 38 (6 March 1818), p. 877.

14 As recalled in Edwin Dunkin *A Far off Vision: A Cornishman at the Greenwich Observatory*, eds Peter Hingley and Tamsin Daniel (Truro: Royal Institution of Cornwall, 1999), p. 45.

Case study D

1 Longitude Act of 1714, CUL RGO 14/1, ff. 10–12 http://cudl.lib.cam.ac.uk/view/MS-RGO-00014-00001/19.

2 Confirmed minutes of the Board of Longitude, 9 February 1765, CUL RGO 14/5 http://cudl.lib.cam.ac.uk/view/MS-RGO-00014-00005/79.

3 Maskelyne to Lord Sandwich, 22 April 1773, NMM SAN/F/22:2 http://cudl.lib.cam.ac.uk/view/MS-SAN-F-00004/2.

4 Howse, *The Seaman's Astronomer*, p. 84.

5 See Howse, *The Seaman's Astronomer*, p. 256, n. 17.

CHAPTER 5

1 Extracts from the Royal Society Confirmed Minutes, 3 July 1760.

2 Nevil Maskelyne, 'Concerning the Latitude and Longitude of the Royal Observatory at Greenwich', *Philosophical Transactions,* 77 (1787), 151–87.

3 James Bradley, 'A Letter to the Right Honourable George Earl of Macclesfield concerning an apparent motion Observed in Some of the Fixed Stars', *Philosophical Transactions*, 45 (1748), 1–43.

4 Apprenticed to Henry Stanbury, freed 16 January 1720 – The Company of Clockmakers Register of Apprentices 1631–1931.

5 John Ellicott, 'The Description and Manner of Using an Instrument for Measuring the Degrees of the Expansion of Metals by Heat', *Philosophical Transactions,* 39 (1735), 297–9.

6 Nevil Maskelyne, 'Observations on a Clock of Mr Shelton, made at St. Helena', *Philosophical Transactions,* 52 (1761), 434–43.

7 James Bradley, 'An Account of Some Observations made in London, by Mr George Graham, F.R.S. and at Black River in Jamaica, by Colin Campbell, Esq. Concerning the Going of a Clock', *Philosophical Transactions,* 38 (1733), 302–14.

8 Nevil Maskelyne, 'An Account of Observations Made on the Mountain Schehallien for Finding its Attraction', *Philosophical Transactions,* 65 (1775), 500–42.

9 Nevil Masklelyne to Edmund Maskelyne, 23 December 1763, NMM MP http://cudl.lib.cam.ac.uk/view/MS-REG-00009-00037/315-317.

10 NMM microfilm of Harrison's Journal, Corry copy, p. 104.

11 Nevil Maskelyne, *An Account of the Going of Mr. John Harrison's Watch, at the Royal Observatory, from May 6th, 1766, to March 4th, 1767* (London, 1767), p. 24 http://cudl.lib.cam.ac.uk/view/PR-PBD-06075/32.

12 John Harrison, *A Description Concerning Such Mechanism*, footnote 22 http://www.bhi.co.uk/sites/default/files/CSM.pdf.

13 NMM CSR/1-27, Sales from January 1937, p. 50.

14 Erroneously listed in Board of Longitude minutes as number 39.

15 Thomas Mudge, *A Narrative of Facts Relating to Some Time-keepers, Constructed by Mr. Thomas Mudge* (London, 1792), p. 5 http://cudl.lib.cam.ac.uk/view/PR-PBA-01784/23.

16 Nevil Maskelyne, *Astronomical Observations... from the Year MDCCLXXXVII to the Year MDCCXCVIII*, vol. III (London, 1799), p. 339 (end of 1795).

17 Papers on payment for Board of Longitude work, CUL RGO 14/17, p. 185 http://cudl.lib.cam.ac.uk/view/MS-RGO-00014-00017/343.

18 William Coombe to the Commissioners of Longitude, 1783 or 1784, CUL RGO 14/23, p. 221 http://cudl.lib.cam.ac.uk/view/MS-RGO-00014-00023/469.

19 Ibid.

20 Thomas Earnshaw, *Longitude: An appeal to the Public* (London, 1808), p. 16.

21 Ibid., p. 40.

22 Ibid.

23 William Hardy, 'Detached Clock Escapement', *Transactions of the Society of Arts*, vol. 38 (1821), p. 81.

24 Charles Frodsham, 'Dead-Beat Escapements', *The Horological Journal* (March 1886), p. 108.

25 Maskelyne, *An Account of the Going of Mr. John Harrison's Watch*, p. 24.

26 WRO 1390/2G, Memoranda 1791–6, p. 2.

27 WRO 1390/2H, Memoranda 1797–1801, p. 3.

28 Nevil Maskelyne, Books on chronometer trials, CUL RGO 4/312 http://cudl.lib.cam.ac.uk/view/MS-RGO-00004-00312/29.

29 *Transactions of the Society of Arts*, 38 (1821), 167–8.

Case study E

1 James Cook, *Captain Cook's Journal During His First Voyage Round the World*, ed. W.J.L. Wharton [1893] Project Gutenberg EBook, 2005, http://www.gutenberg.org/files/8106/8106-h/8106-h.htm.

2 Cited in Board of Longitude Minutes, 2 November 1771, CUL RGO 14/5 http://cudl.lib.cam.ac.uk/view/MS-RGO-00014-00005/211.

3 Board of Longitude Minutes, 2 March 1776, CUL RGO 14/5 http://cudl.lib.cam.ac.uk/view/MS-RGO-00014-00005/303.

4 Maskelyne's list of instruments for Cook's third voyage, NMM AGC/8/39 http://cudl.lib.cam.ac.uk/view/MS-AGC-00008-00029/5.

5 Letters, memoranda and journal containing the history of Mr William Gooch, CUL Ms.Mm.6.48, f. 26v http://cudl.lib.cam.ac.uk/view/MS-MM-00006-00048/54.

6 William Gooch, letter to his mother, 29 April 1791, CUL Ms.Mm.6.48, f. 31v http://cudl.lib.cam.ac.uk/view/MS-MM-00006-00048/64.

7 Nevil Maskelyne to William Gooch (senior), 25 January 1794, CUL Ms.Mm.6/48, f. 118r http://cudl.lib.cam.ac.uk/view/MS-MM-00006-00048/237.

8 William Wales, *The Original Astronomical Observations, Made in the Course of a Voyage towards the South Pole, and Around the World* (London, 1777), p. lv.

CHAPTER 6

1 For more information on the foundation of the Commissioners of Longitude and their changing nature and activities over the ensuing decades, see Alexi Baker, 'The Board of Longitude, 1714–1774: The self-fashioning of an early science funding body', in progress at the time of writing.

2 *London Evening Post* (30 June 1737).

3 For more information, see Alexi Baker, 'Jane Squire and the longitude: Gender and religion in early modern science', article in progress at the time of writing.

4 The British Library holds one of these proposals, notated in what is almost certainly the author's hand. Jane Squire, *A Proposal to Determine our Longitude* with MS additions (London, 1731), BL 1397.d. 43.

5 Jane Squire, *A Proposal to Determine our Longitude*, 2nd ed. (London, 1743) at the Museum of the History of Science at Oxford, pp. 26–36 (in the 'Letters').

6 BL IOR/E/1/21 ff. 263–263v, Letter 145.

7 Squire, *A Proposal to Determine our Longitude*, pp. 21–23 (in the 'Letters').

8 Katy Barrett, '"Explaining" Themselves: the Barrington Papers, the Board of Longitude, and the Fate of John Harrison', *Notes and Records of the Royal Society*, 65 (2011), 145–62, p. 8. The note is at the NMM, Barrington Papers, p. 10:1 http://cudl.lib.cam.ac.uk/view/MS-BGN-00000/37.

9 '*C'est par là qu'il a mérité d'être pendant quarante ans le chef et comme le régulateur des astronomes.*' *Mémoires de la classe des sciences mathématiques et physiques de l'Institut de France* (Paris, 1812), p. ixx.

10 William Fuller to the Commissioners of Longitude, 6 March 1783, CUL RGO 14/36:130v http://cudl.lib.cam.ac.uk/view/MS-RGO-00014-00036/256.

11 Royal Society PP1/LBC.20.204, p. 235.

12 Naomi Tadmor, *Family & Friends in Eighteenth-Century England: Household, Kinship, and Patronage* (Cambridge University Press: Cambridge, 2001), p. 167.

13 Miscellaneous correspondence, CUL RGO 4/187 http://cudl.lib.cam.ac.uk/view/MS-RGO-00004-00187/1.

14 Draft letter from Nevil Maskelyne to Lord Sandwich in 1773, RGO 4/187, 48:1r http://cudl.lib.cam.ac.uk/view/MS-RGO-00004-00187/163.

15 Thereza Storey Maskelyne, 'Nevil Maskelyne, D.D., F.R.S., Astronomer Royal' in *Wiltshire Archaeological and Natural History Magazine*, 29 (1897), pp. 126–37, 131.

16 '*Chez lequel je trouvai une politesse et une complaisance que les savants de ce rang n'ont pas toujours pour les passants*', quoted in the notice of Maskelyne's death in *The Gentleman's Magazine* 109 (1811), p. 197.

17 Letters from Maskelyne to Andrews, CUL RGO 4/149 http://cudl.lib.cam.ac.uk/view/MS-RGO-00004-00149/1.

18 *The Gentleman's Magazine*, 109 (1811), p. 197.

Case study F

1 See John Gascoigne, 'The Royal Society and the Emergence of Science as an Instrument of State Policy', *British Journal for the History of Science*, 32 (1999), 171–84.

2 Joseph Banks to Charles Jenkinson (Lord Hawkesbury), 18 November 1788, BL Add. MS 38223, f. 273–74, quoted in ibid, p. 174.

3 David Philip Miller, 'The "Hardwicke Circle": The Whig Supremacy and its Demise in the 18th-century Royal Society', *Notes and Records of the Royal Society*, 52 (1998), 73–91, pp. 80–81. Miller named the group after Philip Yorke, the 2nd Earl of Harwicke.

4 Anon., *An History of the Instances of the Exclusion from the Royal Society ... by Some Members in the Minority* (London, 1784), p. 17.

5 S. Horsley et al., *An Authentic Narrative of the Dissensions and Debates in the Royal Society, Containing the Speeches at Large of Dr. Horsley, Dr. Maskelyne, Mr. Maseres, Mr. Poore, Mr. Glennie, Mr. Watson and Mr. Maty* (London, 1784), p. 66.

6 Michael Lort, antiquarian and chaplain at Lambeth Palace, quoted in C.R. Weld, *History of the Royal Society, with Memoirs of the Presidents*, 2 vols (London, 1848), vol. 2, p. 169.

7 Horsely, *An Authentic Narrative*, pp. 150–51.

CHAPTER 7

1 Richard Holmes, *The Age of Wonder: How the Romantic Generation Discovered the Beauty and Terror of Science* (London: Harper Press, 2009), p. 58.

2 Lynn B. Glyn, 'Israel Lyons: A Short but Starry Career. The Life of an Eighteenth-Century Jewish Botanist and Astronomer', *Notes and Records of the Royal Society of London*, 56. (2002), 275–305.

3 Royal Society Library, RSC/1/1/1, Dining Club Minute Books, 5 May 1768.

4 Joseph Banks, *Endeavour* Journal, State Library of New South Wales, Safe 1/12–13 (entry for 3 May 1771).

5 Maskelyne to Banks, 10 October 1775, NMM MP http://cudl.lib.cam.ac.uk/view/MS-REG-00009-00037/360.

6 Minute Book of 'The Royal Society Club', 23 November 1775, British Library Add. Ms 32445.

7 Ibid.

8 Minute Book of 'The Royal Society Club', 8 May 1777.

9 See H.C. Cameron, *Joseph Banks, the Autocrat of the Philosophers* (London: Batchworth Press, 1952) and John Gascoigne, *Joseph Banks and the English Enlightenment: Useful Knowledge and Polite Culture* (Cambridge: Cambridge University Press, 1994), p. 16.

10 Blagden to Banks, 30 October 1783, published in *The Scientific Correspondence of Sir Joseph Banks, 1765–1820*, ed. Neil Chambers (London: Pickering & Chatto, 2007) [hereafter 'Chambers'], letter [424].

11 Banks to Blagden, 9 November 1782, Chambers [300].

12 Banks to Blagden, 20 October 1783, Chambers [416].

13 See Blagden to Banks, 14 October 1783, Chambers [406] and Banks to Blagden, 18 October 1783, Chambers [412].

14 Draft notes made by Joseph Banks, Royal Society, MM/1/46a.

15 Royal Society, MM/7/41. Draft minutes, in Banks's handwriting, of the Board of Longitude meeting. Messrs Wright and Gill were the partners in an eminent wholesale stationery company. Banks began his draft copy of the minutes with: 'The Astronomer Royal, without the consent of the Committee certified Mr Bonds and Messrs Wright and Gill. The PRS signified his wish to be excused from acting any more on the Committee.'

16 Banks to Blagden, 6 March 1784, Chambers [476].

17 Maskelyne to Banks, 31 August 1784, NMM, GAB 10.

18 Royal Society, Journal Book, JBO/32, 30 November 1784.

19 Maskelyne to Banks, 4 March 1785, NMM MP http://cudl.lib.cam.ac.uk/view/MS-REG-00009-00037/383.

20 Blagden to Banks, 30 September 1785, Chambers [605]; Blagden to Banks, 23 October 1785, Chambers [612]; Blagden to Banks, 30 October 1785, Chambers [614].

21 Blagden to Banks, 22 October 1786, Chambers [694].

22 Maskelyne to Banks, 8 November 1786, NMM MP http://cudl.lib.cam.ac.uk/view/MS-REG-00009-00037/437.

23 Blagden to Banks, including note by Banks, 30 January 1789, Chambers [896].

24 William Roy to Maskelyne, 11 December 1786, Royal Society, DM/4/14.

25 Blagden to Banks, 30 August 1787, Chambers [767].

26 Banks to Mudge Jnr, 13 March 1793, Chambers [1170].

27 Banks to Maskelyne, 10 July 1793, Chambers [1210].

28 Maskelyne to Banks, 15 September 1795, Chambers [1327].

29 Maskelyne to Banks, 28 March 1798, NMM MP http://cudl.lib.cam.ac.uk/view/MS-REG-00009-00037/417.

30 Minutes of the Board of Longitude, 3 March 1803, CUL RGO 14/7 http://cudl.lib.cam.ac.uk/view/MS-RGO-00014-00007/183; Memorandum by Sir Joseph Banks on the question of Mr Earnshaw's chronometers, Royal Society, MM/8/41.

31 Memorandum by Sir Joseph Banks, RS MM/8/41.

32 Maskelyne to Banks, 24 February 1806, Chambers [1851].

33 Banks to Maskelyne, 24 February 1806, Chambers [1852].

34 Maskelyne to Banks, 4 February 1808, State Library NSW, Series 66.14 http://www2.sl.nsw.gov.au/banks/series_66/66_14.cfm.

35 Maskelyne to Banks, 26 May 1809, Chambers [1908].

36 Banks to Maskelyne, 8 January 1810, Chambers [1919].

37 Andrew Story to Revd J. Prower, 15 March 1816, NMM MP http://cudl.lib.cam.ac.uk/view/MS-REG-00009-00037/1175.

38 Banks to Mrs Sophia and Miss Margaret Maskelyne, 11 February 1811, NMM MP http://cudl.lib.cam.ac.uk/view/MS-REG-00009-00037/586.

Case study G

1 Quoted in Mary Arnold-Foster, *Basset Down: An Old Country House* (1950), p. 42.

2 Quoted in Derek Howse, *Nevil Maskelyne: The Seaman's Astronomer* (Cambridge: Cambridge University Press, 1989), p. 196.

3 Arnold-Forster, *Basset Down*, 1950, p. 36. The original portrait by Dance is now at Powis Castle (National Trust).

4 David Kinnebrook to his father, 15 October 1794, CUL RGO 35/101.

5 The bust is likely to be the terracotta by Roubiliac (NMM ZBA1640), modelled in 1731 from a death mask of Newton and bequeathed for display at the Observatory by John Belchier FRS in 1785.

6 This image was kept among the Maskelyne family papers, NMM MP http://cudl.lib.cam.ac.uk/view/MS-REG-00009-00037/1207.

7 See Richard Dunn, 'An Ovidian tribute to Nevil Maskelyne', Longitude Project Blog (8 July 2011) http://blogs.rmg.co.uk/longitude/2011/07/08/ovid/.

CHAPTER 8

1 Maskelyne notebooks held in Wiltshire and Swindon Archives, WRO1390/2F 1788–91.

2 WRO1390/2H, 1797–1801.

3 Ibid.

4 WRO1390/2F 1788–91.

5 WRO1390/2F 1788–91 and WRO1390/2G 1791–6.

6 WRO1390/2F 1788–91.

7 Ibid.

8 Tragacanth is a gum derived from the resin of the goat thorn tree which is common to the eastern Mediterranean. It is used to treat both diarrhoea and constipation, however, taken in conjunction with castor oil, would imply that Mrs Maskelyne suffered from the latter.

9 Jim Caslaw, Purton Museum http://www.purtonmuseum.com/exhibits/2011-10text.htm.

10 'Died – Tuesday Evening, in St. John's Square, the Lady of the Rev. Sir George Booth, Bart.', *St James's Chronicle, or the British Evening Post* (30 March 1784).

11 Guildhall, St Andrew Holborn, Register of marriages by licence, 1781–90, P69/AND2/A/01/Ms 6671/6.

12 Ibid.

13 Lisa Zunshine, *Bastards and Foundlings: Illegitimacy in Eighteenth-century England* (Ohio: Ohio State University, 2005), p. 135.

14 An extensive discussion regarding the reuse of clothing and particularly the Snowshill Collection may be found in Jane Ashelford, *The Art of Dress: Clothes and Society, 1500–1914* (New York: Harry N. Abrams, Inc, 1996), pp. 133–57.

15 Fanny Burney, *Evelina: Or, The History of a Young Lady's Entrance into the World,* vol. 2 (London: T. and W. Lowndes, 1784), p. 61.

16 Throughout the 1770s and early 1780s, Sir George Booth is reported as a key figure in Sunday charity sermons. A London paper, *The Public Advertiser*, regularly records his appearances from February 1773 until 1781.

17 WRO1390/2D.

18 Sophia's portrait is now in a private collection, Nevil's in the Royal Society's.

19 Notebook of accounts of the Reverend Nevil Maskelyne, NMM MP http://cudl.lib.cam.ac.uk/view/MS-REG-00009-00037/2.

20 Quoted in Derek Howse, *Nevil Maskelyne: The Seaman's Astronomer* (Cambridge: Cambridge University Press, 1989) p. 145.

21 See Amanda Vickery, *The Gentleman's Daughter* (Yale: Yale University Press, 2003).

22 NMM MP http://cudl.lib.cam.ac.uk/view/MS-REG-00009-00037/2.

23 Nineteenth century fashion plate collection, Scripps College, Ella Strong Denison Library, Macpherson Collection, box 2.

24 Steven J. Gores, *Psychosocial Spaces: Verbal and Visual Readings of British Culture, 1750–1820* (Wayne State University, 2003), p. 147.

25 Howse, *The Seaman's Astronomer*, p. 196.

26 See Claire Hughes, *Dressed in Fiction* (London: Berg, 2006).

27 This is very well shown in Russell's superb portrait of the young Mrs William Pierrepont, wife of a wealthy naval captain, also in the NMM collection (PAJ2906).

28 Hughes, *Dressed in Fiction*, p. 38.

29 Jane Austen, *Northanger Abbey and Persuasion* (London: John Murray, 1818), p. 17

30 Private collection.

31 Howse, *The Seaman's Astronomer*, p. 206.

32 Bridget Storey to Rev. J. Power, 31 March [n.d.], NMM MP http://cudl.lib.cam.ac.uk/view/MS-REG-00009-00037/1169.

33 Law Intelligence, *The Bury and Norwich Post*, Bury St. Edmunds, Wednesday 12 June 1816.

34 *The Times* (London, England), Thursday 15 February 1821.

35 Alistair Owens, 'Property, gender and the life course: inheritance and family welfare provision in early nineteenth-century England', *Social History*, 26 (2001): 299–317, p. 317.

CODA

1 *'Vitæ doctissimæ Utilitatem'*. The full Latin inscription, and a translation by Lesley Murdin, are in Derek Howse, *Nevil Maskelyne: The Seaman's Astronomer* (Cambridge: Cambridge University Press, 1989), p. 235.

2 See Robert W. Smith, 'A National Observatory Transformed: Greenwich in the Nineteenth Century', *Journal for the History of Astronomy*, 22 (1991), 5–20.

3 This was said in support of an application for a pension by Pond's widow: G.B. Airy to Henry Warburton, quoted in George Biddell Airy, *Autobiography* (Project Gutenberg, 2004), p. 71 http://library.mat.uniroma1.it/appoggio/MOSTRA2006/airy_biography.pdf.

4 [Patrick Kelly], 'Maskelyne, Nevil', in *The Cyclopaedia; or, Universal Dictionary of Arts, Sciences, and Literature*, ed. by Abraham Rees, vol. 22 (London, 1812), pp. 416–22, p. 422.

5 Vitam bene actam.

6 Howse, *The Seaman's Astronomer*, p. 210.

7 Samuel Vince to Margaret Maskelyne, 8 July 1812, NMM MP http://cudl.lib.cam.ac.uk/view/MS-REG-00009-00037/630.

8 Howse, *The Seaman's Astronomer*, p. 235.

ACKNOWLEDGEMENTS

My thanks go to those who helped to make the 2011 Maskelyne Symposium at the National Maritime Museum (NMM) a success. These include the speakers: Jim Bennett, Mary Croarken and Amy Miller, whose talks became chapters in this book; Jenny Gaschke, some of whose insights fed into my analysis of the portraits and drawings in the Maskelyne Collection; and Richard Dunn and Rory McEvoy, who spoke about objects and manuscripts from the collection – the last of which also became a chapter. I am grateful for the support of the NMM's Research and Curatorial Department, especially Lizelle de Jager, Gloria Clifton and Nigel Rigby.

The book has benefited from additional contributions beyond those presented in 2011. Thanks to Caitlin Homes for her chapter and invaluable catalogue of the Maskelyne Papers, arising from her research internship at the Museum. Alexi Baker's and Nicky Reeves's chapters are two of many outputs associated with the joint NMM/University of Cambridge project, funded by the AHRC, on the history of the Board of Longitude. This project and the Jisc-funded digitisation of the Board's and Maskelyne's papers has provided a supportive and stimulating background to my own contributions: thanks to Richard, Alexi, Nicky, Simon Schaffer, Katy Barrett, Eoin Philips, Sophie Waring, Huw Jones, Lucinda Blaser, Mary Ferguson, Megan Barford and James Poskett.

Final thanks to Kara Green, for all her work on the drafts, proofs and images, Rebecca Nuotio, the photo studio at the NMM, and Nikki Edwards at Robert Hale.

Rebekah Higgitt

INDEX